WORKING WITH CONGRESS

A PRACTICAL GUIDE FOR SCIENTISTS AND ENGINEERS

William G. Wells, Jr.

THE AMERICAN ASSOCIATION
FOR THE ADVANCEMENT OF SCIENCE
AND THE
CARNEGIE COMMISSION ON SCIENCE, TECHNOLOGY, AND GOVERNMENT

Library of Congress Cataloging-in-Publication Data

Wells, William G., Jr.
Working with Congress: a practical guide for scientists and engineers /
prepared by William G. Wells, Jr. for the American Association for
Advancement of Science and the Carnegie Commission on Science,
Technology, and Government.
        p.   cm.
   Includes bibliographic references.
   ISBN 0-87168-504-3
   1. Scientists—United States—Handbooks, manuals, etc.
2. Engineers—United States—Handbooks, manuals, etc.   3. Science and
state—United States—Handbooks, manuals, etc.   4. Engineering and
state—United States—Handbooks, manuals, etc.   5. United States.
Congress—Handbooks, manuals, etc.   I. Title.
Q149.U5W46   1992
338.97306—dc20   92–35313
                                                           CIP

The AAAS Board of Directors, in accordance with Association policy, has
approved the publication of this work as a contribution to the
understanding of an important area. Any interpretations and conclusions
are those of the author and do not necessarily represent views of the Board
or the Council of the Association.

Design and illustration by Julie A. Cherry

AAAS Publication: 92-31S
International Standard Book Number: 0-87168-504-3

Printed in the United States of America

✪
Printed on recycled paper

Contents

FOREWORD

Much has been said and written about what C.P. Snow has described as the "two cultures" and the gulf that seems to separate the scientist from society and those who govern it. Although one can certainly point to differences between scientists and elected officials, they have much in common, and important forces bring them together. Science works for society, and society in turn depends on and nurtures science. Both scientists and elected officials share a mandate to work in the interest of the nation. As Albert Einstein once said: "Concern for man himself and his fate must always form the chief interest of all technical endeavors. . . . Never forget this in the midst of your diagrams and equations."

As William G. Wells, author of this guide, points out, all citizens have the right to let their elected representatives know their views on issues that concern them. This guide offers scientists and engineers clear and detailed advice on how to communicate with senators, representatives, and their staffs.

For communication is the key to bridging the gap between the mutually dependent "cultures" of science and government. *Working With Congress* is intended to give scientists and engineers a glimpse of the inner workings of the legislative branch of the federal government. As events of recent years have demonstrated, Congress, as the source of the nation's laws and the appropriator of federal funds, exerts considerable influence over the future directions of science and technology. It is in the interest of scientists and engineers to develop a basic understanding of congressional operations in order to communicate effectively with the 535 elected senators and representatives who constitute the legislative branch.

The Carnegie Commission on Science, Technology, and Government is very pleased to have sponsored the production of this important publication. We congratulate the American Association for the Advancement of Science, Dr. Wells, and others who contributed to the development of *Working With Congress*, and we hope that the scientists and engineers to whom it is directed will find it useful as a practical guide to the legislative branch of our government.

John Brademas

John Brademas
Chair, Committee on Science,
Technology and Congress

William T. Golden

William T. Golden

Joshua Lederberg

Joshua Lederberg

Co-Chairs, Carnegie Commission on Science,
Technology, and Government

PREFACE

The agenda of the United States Congress is increasingly dominated by issues involving science and technology. Responding effectively to critical and complex challenges such as economic competitiveness, global environmental change, the AIDS epidemic, energy uncertainties, and national security requires that Congress have access to the best scientific and technical information and advice. At the same time, the scientific and technological community is ever more reliant on government support and thus on Congress.

Few members of Congress have any training or background in science or engineering. Although the number of staff with technical credentials has grown substantially in recent years, it is still small. Many scientists have contact with the executive branch of the government through grant support, participation on advisory committees, or other mechanisms, but only a handful have much experience in dealing with Congress. Many observers have noted that scientists and engineers have not been as effective as they might be in their interactions with Congress and have suggested that improved communication could benefit both the research community and the quality of national policymaking on science and technology issues.

The American Association for the Advancement of Science (AAAS) and the Carnegie Commission on Science, Technology, and Government have a common interest in improving the relationships between science, technology, and government. AAAS, founded in 1848, is an interdisciplinary organization of some 135,000 members. Through its Committee on

Science, Engineering, and Public Policy, it conducts programs to improve communication between scientists and policymakers, to foster improved public policy in areas related to science and technology, and to enhance the capacity of the science and technology community to contribute to national goals. The Carnegie Commission was created in 1988 by the Carnegie Corporation of New York to examine the processes through which government incorporates scientific and technological knowledge into the formulation and implementation of policy. The commission's focus is on the organization and decisionmaking processes of government. Its Committee on Science, Technology, and Congress produced two reports in 1991 outlining specific recommendations for both Congress and the scientific and engineering community to improve their interaction regarding science and technology issues.* One key recommendation was that more scientists, engineers, and others should become actively involved in public policy activities, especially in relation to Congress.

AAAS has developed this guide under sponsorship of the Carnegie Commission with the aim of enabling individuals in the scientific and engineering community to understand, work with, and communicate more effectively with the U.S. Congress. The guide describes the constitutional basis of Congress, its culture and traditions, its power structure and form of organization, and its principal activities. Its major purpose, however, is to serve as a practical manual to help scientists and engineers in working with Congress — through personal visits, telephone interactions, correspondence, and participation in hearings.

It is our hope that this book will encourage scientists and engineers to become more involved in congressional activities and that it will improve the effectiveness of those who choose to do so. Those who deplore the low level of understanding of science and technology among political leaders should recognize that this situation has a "flip side." To quote from the

---

* *Science, Technology, and Congress: Expert Advice and the Decision-Making Process*, Carnegie Commission on Science, Technology and Government, (Feb. 1991); *Science, Technology, and Congress: Analysis and Advice from the Congressional Support Agencies*, Carnegie Commission on Science, Technology and Government (Oct. 1991).

concluding chapter of the guide, "If we, as scientists and engineers, expect Congress to understand us, it is essential that we make more of an effort to understand and work with them." Finally, the personal dimension should not be overlooked. Congressional involvement can be highly rewarding for the individuals concerned. There is satisfaction to be gained from seeing one's ideas influence policy and one's expertise used to benefit the nation.

The author of *Working With Congress*, Dr. William G. Wells, Jr., is an engineer who has a wide range of experience in science and government — as an Air Force officer, as a manager in NASA's Apollo Program, as a congressional staff member (including many years as a staff director for key science subcommittees in the House of Representatives), and as chief of staff at the White House Office of Science and Technology Policy. He served as head of the Office of Public Sector Programs at AAAS from 1980 to 1984 and is currently professor of management science at George Washington University in Washington, D.C. In addition to drawing on his own experience and a variety of published sources in preparing this guide, Dr. Wells conducted a questionnaire and interview survey of members of Congress and congressional staff to elicit their views on interactions with scientists and engineers. The results of this survey are incorporated both in the guidance contained in Chapters 4 and 5 and in a number of boxed quotes that appear throughout the text.

On behalf of AAAS, we would like to express our deep appreciation to Dr. Wells for carrying through this project amidst his numerous other activities. In addition, we are grateful to the Carnegie Commission on Science, Technology, and Government for its financial support of this project. Many individuals contributed to this volume. The members of Congress and congressional staff who responded to Dr. Wells's survey deserve special thanks. Many of these individuals also took the time to comment on drafts of the document. Their comments were extremely useful and we appreciate them. Others who commented on drafts and to whom we are grateful include Jesse Ausubel, Richard Barke, John Brademas, Rick Brandon, Lynne P. Brown, Kathryn Edmundson, B. R. Inman, Pamela Kulik, David Robinson, and Mark Schaefer. We are especially grateful to Jeannette Aspden for editing the final draft. Thanks are due to Mary E. Morrison who assisted Dr. Wells in the early stages of this project, and to the AAAS Publications Office who transformed the manuscript into this book. Finally, we are also grateful to the members of an *ad hoc* subcommittee of the AAAS Committee on Science, Engineering, and Public Policy who provided overall guidance to the effort and also reviewed the manuscript at various stages.

Additional copies of this book are available from AAAS. Comments and suggestions regarding this book are welcome and may be addressed either to AAAS or the Carnegie Commission.

Albert H. Teich and **Stephen D. Nelson**
Directorate for Science and Policy Programs
American Association for the Advancement of Science

Washington, D.C.
November 1992

American Association for the Advancement of Science

Committee on Science, Engineering, and Public Policy

Ad Hoc Subcommittee on Guide to Congress Project

John Andelin
Assistant Director
Office of Technology Assessment
U.S. Congress

Dennis Barnes
President — Southeastern
Universities Research Association

Nancy Carson
Program Manager — Science,
Education & Transportation
Office of Technology Assessment
U.S. Congress

Christopher T. Hill
Principal Program Officer
National Academy of Engineering

Thomas H. Moss
Dean of Graduate Studies
and Research
Case Western Reserve
University

Lesley Russell
Professional Staff Member
Committee on Energy
and Commerce
U.S. House of Representatives

Michael L. Telson
Senior Analyst
Committee on the Budget
U.S. House of Representatives

CHAPTER 1

WHY WORK WITH CONGRESS?

Why should individual scientists and engineers become involved and work with Congress? Congressional activities take time, and in today's highly competitive technical environment, most scientists and engineers have barely enough time to do their research, pursue funding to keep it going, and fulfill their teaching, administrative, and professional responsibilities. Furthermore, many technically trained people are put off by the rhetoric and apparent irrationality of much of what goes on in Congress. They are more comfortable in the laboratory or out in the field than in the highly charged political atmosphere of Capitol Hill. Then, too, they are skeptical of having any meaningful impact in competition with so many professional lobbyists and more powerful interests.

Most researchers have no such reservations about dealing with the executive branch of the federal government. In fact, it is a rare scientist or engineer who does not have some connection with one or more federal agencies. Academics might be involved with the National Science Foundation or the National Institutes of Health, either as a recipient of grant, contract, or fellowship support; as a member of a review or advisory panel; or as a colleague or collaborator with a government researcher. Comparable relationships often exist between researchers in the private sector and in nonprofit scientific and technical entities and other government agencies including NASA, the Departments of Defense and Energy, and the Environmental Protection Agency.

Half a century ago, the question of whether science should become more deeply involved with government was hotly debated in meetings of the National Academy of Sciences and professional associations such as the American Association for the Advancement of Science. Ultimately, these debates of the 1930s and 1940s were resolved by the accession of leaders — in government as well as science and engineering — who understood that a threshold had been crossed by the middle of the twentieth century. Beyond that threshold, the interweaving of science, technology, and government has become inevitable and increasingly important to both sides.

With the growing linkages between science, technology, and government, scientists and engineers have important, direct reasons for being in frequent contact with executive branch agencies. But Congress is different. Its work goes on at the macro level, in setting policy directions or providing funds for programs which are carried out by executive branch agencies. In general, Congress does not provide funds to individual institutions or research projects.* Thus, few scientists and engineers have immediate needs to work with Congress and relatively few do so regularly.

This guide is intended to persuade you that working with Congress is more than a casual matter for scientists and engineers — it is essential today and will be even more vital in the future. The book also is intended to give you the tools and understanding that can increase your effectiveness on Capitol Hill and can make your experience a positive and productive one.

There are many reasons to work with Congress, but in general they can grouped under three headings: (1) serving the public and national interest; (2) serving the interests of science

---

* The increasingly frequent exceptions to this, known as "pork barrel" projects, have been the subject of considerable discussion in recent years.

2

'In layman's terms? I'm afraid I don't know any layman's terms.'

NICK DOWNES

and engineering and their component disciplines; and (3) serving one's own personal or institutional self-interest. Scientists and engineers can serve the public and national interest by providing advice, information, and opinions based on both special scientific and technical expertise and general understanding of science and technology. The applications of science and technology *to* policy — for example, the scientific understanding of global climate change — are much broader than the questions that involve policy *for* science and technology — such as the funding levels for various scientific and technical disciplines and agencies. Scientists and engineers can contribute in both arenas. They can offer views and information relating to government programs, regulations, and other policy issues based on their own research and special expertise or, in some instances, on the basis of their general scientific and technical knowledge. Such views are seldom "value free." Of necessity, they reflect judgements that go beyond purely scientific considerations and into the realm of values. Political leaders, including members of Congress, are accustomed to dealing with advice in these terms, but scientists and engineers who offer it should still strive to make clear when their ideas are based on generally accepted data and interpretations and when they are more speculative or are personal opinions or judgments.

Providing advice on issues involving science and technology on an altruistic basis is an activity with evident appeal. But it is important to recognize that self-interest, both direct and indirect, on the part of those in the scientific and engineering communities, is also a respectable motivation. Scientists and engineers in all fields have an obvious interest in the welfare of the research enterprise as well as their discipline and research areas and their levels of federal support. There is no reason to be shy about pursuing this kind of indirect self-interest by acting as an advocate for science and engineering or one's own field. Carried out in an appropriate, responsible manner, such advocacy is normal, respectable behavior in the political environment.

In terms of direct self-interest, virtually all scientists and engineers in universities and government and a great many in industry and other institutions are profoundly affected by federal policies and budgets. A researcher might support increased funding for an area of research because support for his or her own institution, department, or laboratory would stand to benefit from larger appropriations. While most scientists would prefer to couch their arguments in more lofty terms, political leaders *expect* citizens to pursue their own interests, and scientists and engineers need not be hesitant in doing so.

Should scientists and engineers become more involved in working with Congress? The alternative is to leave congressional policymaking in areas of importance to science and technology in the hands of other groups and interests. Decisions on support for science and engineering research and on other policy issues involving scientific and technical considerations will be made regardless of whether scientists and engineers choose to become involved. To ignore Congress or to hold it in disdain and remain aloof is to forego the chance to influence policy and to abdicate one's responsibility to the science and engineering communities — and to the nation.

CHAPTER 2

THE CONGRESSIONAL ARENA

This chapter provides a brief overview of Congress, including its constitutional setting, distinguishing features, and relationships with the other branches of government. The focus is on aspects that should be of interest to those who wish to work with Congress but who are not professional scholars or analysts of Congress.

CONSTITUTIONAL SETTING OF CONGRESS

The United States Constitution is a product of consummately crafted compromises arising from long months of negotiations, posturing, threats, and patient statesmanship at the Constitutional Convention of 1787. Take an hour to browse through it. The Constitution is a remarkable document that represents a stunning intellectual and political achievement. A prime example of "less is more," it is founded on a handful of basic principles: limited government, separation of powers, checks and balances, federalism, and popular control of government.[1]

These principles continue to guide lawmaking and the operations of government today, despite the enormous changes that have transformed the world and the role of government in

American society over the past two hundred years. Consider for a moment the Constitution's durability and continuing relevance in this world of change, and how these principles have served the nation.

LIMITED GOVERNMENT

The nation builders who crafted the Constitution sought to establish a strong and effective national government (after the dismal failure of the Articles of Confederation), but at the same time they wanted to protect personal and property rights by avoiding the concentration of too much power in that government. The limitations and constraints were achieved by dividing power among the three branches of government and between the national government, the states, and the citizens. And the first 10 amendments to the Constitution, the Bill of Rights, brilliantly designed to ensure the rights and liberties of the citizens and to protect them against an overreaching government, are still a shining beacon in the affairs of mankind.

SEPARATION OF POWERS

A monopoly of governing power was precluded by establishing three independent branches of government. This separation of powers was designed to limit and constrain the authority of any one branch, but it was also designed to ensure cooperation among the three branches: such cooperation is necessary to make things work in practice.[2] Notwithstanding this overall design, it is clear that the framers of the Constitution had a distinct bias toward representative assemblies. Article I of the Constitution shows a sharp, central focus on the powers of Congress. Not only did the framers of the Constitution identify these powers at length, but they also named Congress as the first branch of government and, in case they missed something, granted Congress a wide range of explicit and implied powers via the so-called elastic clause (Article I, Section 8).

Quite in contrast, Articles II and III, creating the executive and judicial branches of the national government, describe rather briefly the framework and duties of these branches. In short, these Articles are vague and lean compared to the treatment of Congress.

The basic nature of the presidency, described in Article II, was not then, nor has it yet been, clearly fixed. This is with good reason, according to various constitutional scholars. The Constitutional Convention was sharply divided on the issue of executive power; ultimately the participants settled on a compromise in order to get an agreement. Thus, the Constitution reflects the struggle between two conceptions of executive power: on the one hand, it always ought to be subordinate to the supreme legislative power; on the other hand, it ought to be, within limits, autonomous and self-directing. The practical effect (and very likely the intended plan of adherents of strong executive power) was to plant the seeds of implied presidential power and the long-term development of the presidency as "what the President thinks it is," at least within the limits set by the system of checks and balances.

Article III, which establishes the judiciary, is even leaner than Article II. And here again, Congress has a role, being given the power to establish new judicial positions and create new courts below the Supreme Court. The sparseness of the constitutional language enabled the fourth Chief Justice, John Marshall, to establish the principle of judicial review, providing the federal courts with the power to rule on the constitutionality of specific legislative and executive decisions. Marshall's role illustrates an important dimension of the flexibility of the Constitution and the American political system. Within limits, the nation's affairs can be tremendously affected by a strong and determined individual whether in the legislative, the executive, or the judicial branch.

CHECKS AND BALANCES

Closely connected to the principle of separation of powers is that of checks and balances. Not only did the framers anticipate that strong individuals in each branch might seek to expand their own power at the expense of the other branches, they actually crafted what amounts to an open invitation to conflict and struggles for power, especially between Congress and the President. To contain this conflict and restrain each branch, the framers devised an intricate system of check and balances.

Familiar examples of these checks and balances include the internal constraints within Congress arising from its bicameral nature; the ability of the President to veto laws passed by Congress; and the requirement that budgets, treaties, and high-level appointments proposed by the President must be approved by Congress. In addition, many decisions and actions of Congress and the President are subject to review by the federal judiciary. Finally, judicial decisions — including those of a constitutional nature — may be overturned by Congress and the President working in concert. Such checks and balances have a dual effect: they introduce the potential for conflict, but they ultimately encourage cooperation, bargaining, negotiation, compromise, and accommodation among the branches.

FEDERALISM

Just as the three branches of the national government check each other, the Constitution includes another dimension of checks and balances and division of power between the state and federal governments. The most important aspect of this principle is that powers not granted to the national government by the Constitution remain with the states and with the people.

POPULAR CONTROL OF GOVERNMENT

The framers of the Constitution intended the House of Representatives to be the most representative part of the national government. Representatives are elected directly by the people for two-year terms to ensure that the people's will is continuously reflected within the government. In practice, for many members of the House, this means campaigning nonstop — visiting their districts, making speeches, serving constituents, and raising campaign money — as well as attending to the lawmaking responsibilities of a member of Congress.

Originally, the Senate was designed to be one step removed from popular voting in order to moderate the popular will of the House. State legislatures were given the responsibility of electing senators. Eventually, this process was swept aside with the enactment of the Seventeenth Amendment in 1913; this amendment provided for direct election of senators, thereby more firmly establishing "the electoral connection" — to use David Mayhew's phrase — as a central feature of U.S. governance.[3] And while senators have six-year terms, and thus do not face elections as frequently as House members, they too have become intensely responsive to constituents. Members of Congress are keenly attuned to the power of the people to "turn the rascals out" if they so choose. From time to time, that is exactly what the people do.

HOW MEMBERS VIEW CONGRESS AND THEMSELVES

Aside from this overall constitutional appraisal, it is important to understand something of how members of Congress view themselves and their institution. Members of Congress, along with the President and the Vice President, are acutely aware of belonging to a highly select group that shares an almost mystical bond: they were elected to national office by the citizens of the United States. They never forget this and do not want others to forget it either. Membership in this very special community separates them constitutionally and in their heart of hearts from appointed officials — including even Supreme Court justices and cabinet secretaries — and from staff members, however close such individuals might be professionally or personally. It is a bond that transcends partisan concerns and branches of government.

No matter how powerful congressional staff members may become, it is perilous for them to presume equality with a senator or representative. Sometimes senior individuals from industry or academia, who have come into the executive branch at upper levels, have an encounter with a member of Congress that clearly places the appointee in a status inferior to even a junior member. Even White House staff members, as powerful as many of them believe themselves to be, must tread carefully in dealing with a member of Congress — including those of their own party. Obviously, there are some exceptions to this rule, but the general point remains valid.

However "green" and untutored new members of Congress may be upon arrival in Washington, they soon absorb the implications of their initiation into this select society. Although members of Congress can be as open as the average citizen to flattery and courting by the President and as impressed as anyone by a personal call from the Oval Office or an invitation to dinner at the White House, they nevertheless see themselves as constitutionally coequal with the President in a collective sense. More than a few Presidents (and their staffs) have had to learn this the hard way. Particularly in private, when a President attempts to bully or threaten a member, the member will more often than not respond in kind. Those Presidents who have been most effective in getting their programs passed have tended to treat members of Congress as coequals.

As a constitutional principle, the separation of powers is central to our political system. Just as important, it is built into the very fiber of individual members of Congress as well as the institution collectively. It is a concept that often outweighs party bonds. While a President usually can count on support from members of Congress from his own party, such support is not automatic — as every President learns. Indeed, support will be withheld or given only very grudgingly if it appears that congressional prerogatives are being slighted. If they are seen as being trampled upon, the President can almost certainly expect opposition even from members of his own party. Such behavior is not simply a matter of personal pique; it rests on a deep sense of constitutional integrity.

Finally, a less-than-attractive aspect of membership in this select society sometimes emerges in the form of arrogance and domineering personal behavior. It takes a strong personality to resist the blandishments of life on Capitol Hill. The Hill is organized to cater to the commands and wishes of members of Congress. Perks abound in great variety; staff of all types (personal, support, committee) work furiously to satisfy member requests; the Capitol police stop traffic for members to cross the street; and every facet of congressional operations is geared to support the notion of constitutional uniqueness and privilege of members.

CONGRESSIONAL DUALITY

Roger H. Davidson and Walter J. Oleszek start their excellent analysis, *Congress and Its Members*, with two quotes that express a central thesis about Congress: its fundamental duality.[4]

> A good government implies two things: first, fidelity to the object of government, which is the happiness of the people, secondly, a knowledge of the means by which that object can be best obtained.
>
> *The Federalist*, No. 62 (1788)

> Legislatures are really two objects: a collectivity and an institution. As a collectivity, individual representatives act as receptors, reflecting the needs and wants of constituents. As an institution, the Legislature has to make laws, arriving at some conclusions about what ought to be done about public problems.
>
> Charles O. Jones, "From the Suffrage of the People: An Essay of Support and Worry for Legislatures" (1974)

Davidson and Oleszek use these two statements to capture the essence of the dualism of Congress or in their words, the concept of "two Congresses." This concept, which embraces the idea that representative assemblies contain an inherent tension between lawmaking and representation, is neither especially new nor novel. Rather, as they say, "this dualism is embedded in the Constitution, manifested in history, and validated by scholars' findings."[5] Political scientist Giovanni Sartori, for example, has distinguished between the two functions of Congress by describing them as the "function of functioning" and the "function of mirroring."[6] The dualism is also apparent in the experience of members of Congress who were consulted in preparing this guide.

According to scholars, observers of Congress, and members themselves, one of the "two Congresses" is a lawmaking institution. This is the Congress of most textbooks, the media, and especially the daily network television news. It can be seen as a collegial body, performing constitutional duties and handling legislative issues. This Congress serves as an arena of political combat where many forces of American, and increasingly international, politics converge. These forces are at work within an intricate network of congressional structures, procedures, and personalities that sets the rules and organizes the legislative struggles.[7]

But there is a second Congress, albeit closely bound up with the first and just as important. This is the representative assemblage of 540 senators, representatives, and delegates.[8] Despite widely differing ages, diverse backgrounds, and varying routes taken to office, these men and women share a common experience. Their political lives and electoral fortunes and prospects depend, not so much on what Congress produces as an institution, but upon the support, goodwill, and perceptions of voters in their districts and states. This leads to arrangements that are built on "service" to constituents in all its forms. From this concept arises an important truth that one must grasp if one is to achieve a real understanding of "the Congress." By no means does all congressional activity take place on Capitol Hill.[9] Much of it goes on in the cities, towns, and farms far removed from Washington.

IMPLICATIONS OF CONGRESSIONAL DUALITY

The duality of Congress is manifested first in the schedules of members and the enormous time pressures on those schedules. A member's institutional and individual duties are always there, on today's schedule and months into the future. Not as apparent but always present is the tension between these dual dimensions of the member's schedule. District and state staffs can at times get into a tug of war with Washington staffs over the member's schedule and the priorities it reflects. Virtually every member of Congress would endorse Davidson and

Oleszek's observation: "Senators and Representatives suffer from a lack of time to accomplish what is expected of them. No problem vexes Members more than that of juggling constituency and legislative tasks."[10] There is very little time to read and study, although members are required to make major policy decisions. A comprehensive study by the House Commission on Administrative Review (the Obey Commission) revealed that the average day for a representative exceeded 11 hours and that only a very small part of the day was available for reading or "thinking time." By far, the largest components of time are devoted to meetings in the office and in committees.[11]

CONSTITUENCY BUSINESS — DEMANDS AND OPPORTUNITIES

The average representative spends about 120 days a year in the district, and the average senator 80. Even in Washington, some evidence suggests that members spend less than half of their time on lawmaking and related duties and the balance on constituency activities.[12] There is no doubt that the pull of constituency business is relentless. Three anecdotes point to the very special nature of the constituent-member relationship.

- At midnight, a midwestern representative, accompanied by an aide, wearily dragged himself into a roadside diner for a cup of coffee before heading to the local motel. The day had started at 6 a.m. with a breakfast meeting at a plant 30 miles away. Events had been back-to-back all day. As the coffee arrived, a voice came from a man plunking himself onto the stool next to the representative. "Hello, Congressman. I'm glad to run into you. I wanted to talk to you about a West Point appointment for my nephew." In the member's own words: "The last thing in the world I wanted to do was talk about West Point; I was so damn tired I couldn't think, but I had to turn on the smile and ask my assistant to get out his notebook to take down the information. In many ways this is a 24-hour a day, seven day a week job, but I always remember Harry

Truman's advice: 'If you can't stand the heat, get out of the kitchen.'"

■ A western senator had gone into a bar on a hot, dusty afternoon to cool off with a beer. Barely into his first sip, he noticed a woman getting up from a table and approaching him with a determined look. After placing herself squarely to his right with feet wide apart, she let go with a stream of profanity about how he was not handling his job, mentioning several local issues in particular. Turning on her heel, she returned to her table and friends, leaving the senator to his beer. A few minutes later, she got up again and came over. But this time she said, "Excuse me Senator, but I forgot to ask for your autograph for my son."

■ A midwestern Democratic representative explains the organization of his upcoming district political campaign: "Some of my most ardent supporters and tireless volunteer workers are Republican women whose military sons have been assisted through our constituent service. But we never ask their politics when they need help; I serve all of my constituents. If they decide to reciprocate later, that is up to them." He adds that there is a lot more than constituent service in a campaign but concludes, "A lot of my colleagues have seen the light."

The "light" originates not only in the experiences of members that are passed along to new members but also in the research of congressional scholars like Richard F. Fenno, Jr., who spent nearly a decade visiting and traveling with members in their districts to gain a better understanding of that "other Congress" away from Washington. Fenno, in commenting on a particular representative, said: "For a congressman facing the prospect of close elections in a district opposed to him philosophically, the appeal of a constituency service emphasis would have to be strong. Constituency service is totally nonpartisan and nonideological. As an electoral increment, it is an unadulterated plus." Fenno concludes that constituency service is especially appealing to a representative from a highly heterogeneous district or one that is bewildering from the standpoint of issue politics.[13]

PUBLIC PERCEPTIONS OF INDIVIDUALS AND INSTITUTIONS

The duality of Congress goes a long way toward explaining a well-known paradox in public perceptions of Congress. The lawmaking (institutional) Congress is often seen as messy and incompetent and is held in low esteem by the public, as measured in public opinion polls. The institutional Congress seems to be evaluated by citizens on the basis of their overall attitudes about politics, government policies, and the state of the nation.[14] Also, disapproval of the misbehavior by a small minority of its members tends to spill over and color the public's views on the entire institution.

In contrast, citizens seem to view *their own legislators* as agents of local interests. Citizens measure the performance of their members differently from that of the institution as a whole. Specifically, they place high value on service to the district, communication with constituents, and, using Fenno's phrase, "home style" — the way a member deals with the home folks.[15] The high rate of return of incumbents — from both parties — suggests that citizens throughout much of the nation draw a rather sharp distinction between their own representative and senators and "the rest of those politicians." Some in Congress — and out of it — offer a pungent explanation for people making this distinction: it is argued that much of politics is about "taxing them and spending on us." From this it is argued that "my Congressman brings home the things we need, while all those other bums tax us to pay for 'pork' for those other people."

TWO WORLDS

Of more than passing importance to users of this guide are the great differences between the world of Washington (and the lawmaking Congress) and the world of the district and state (the constituency Congress). Members must adapt to moving back and forth between the environment of Washington, where they are powerful and catered to, and the district environment, where they are seen as servants of local interests. The anecdotes above and a wealth of other data underscore the enormous difference between the relationships members have with constituents compared with the relationships lobbyists, staff, and executive branch officials have with members. This is not to suggest that constituents need never show deference; nor is it to say that all members are predominantly constituency oriented -- some powerful individuals have emphasized their institutional roles. Yet, it should be clear that constituents relate to the members in a unique way that is in harmony with the core, central interests of the member.

IMPLICATIONS FOR STAFFING

Finally, this duality of Congress is reflected in the way members staff their offices. Increasingly, representatives and senators have established a number of offices in their districts and states and have located sizeable fractions of their staff in these offices. For example, former political science professor and now representative from North Carolina, David Price says that, on being elected, "I deployed my staff in a fashion that has become common in the House, setting up several district offices and locating most constituent-service functions there." Price adds that most weeks he is in the district more than in Washington, working in and around three district offices where half his staff is posted. He concludes by saying that "much of what we do in Washington is district-centered as well . . . [including] the work of my legislative assistants, most of whose time is spent dealing with the policy concerns of local groups and correspondents."[16]

OUT OF TOUCH OR OVERLY SENSITIVE?

With the stressful political atmosphere of 1992, there has been a barrage of nonsense from radio talk show hosts that "Congress is out of touch." But, in reality, the problem is *not* that members are insulated within the Beltway and out of touch with their constituents. Rather, one could argue that the problem is just the reverse; the necessity to keep the electoral connection strong forces a certain *hypersensitivity* on the part of legislators to the wishes of the home folks. As former Speaker Tip O'Neill has said, "All politics is local." In the aggregate, this may well collide with reaching "national interest" decisions — such as reducing the budget deficit.

THE NEW CONGRESS: DYNAMIC AND FLUID

REFORMS AND POWER SHIFTS

In seeking to understand Congress, it is essential to grasp the fact that it is a fluid, dynamic institution that can change almost before your eyes. In short, one cannot study it once and go away thinking one knows what one needs to know in order to understand its workings. Lawrence C. Dodd and Bruce I. Oppenheimer in their continuing editions of *Congress Reconsidered* have tracked the unfolding evolution of this complex institution.

During the 1970s and early 1980s, as a result of reforms undertaken in the House in the early 1970s and the Senate in the mid-1970s, Congress was becoming a fragmented, decentralized institution, dominated by subcommittee government. The mid- and late-1980s saw a quite different pattern. There was little discretionary money for new programs and thus little opportunity for policy innovation by the authorizing subcommittees in Congress. Dodd and Oppenheimer suggest that "power has shifted toward the party leadership and a few elite committees that deal with essential annual legislation."[17] Another trend has been the growth of unofficial and quasi-official groups and caucuses to focus attention on particular issues and often to coordinate efforts of those inside and outside Congress.

Despite this power shift, particularly in the House, there seems to be a delicate balance of contending forces. No single locus predominates, as had been the case in earlier decades of committee government or speaker dominance. Power is divided between committees and centralized leadership in such a way as to be highly susceptible to short-term forces. This leads to the prospect of a fluid House of Representatives subject to extensive changes in its governing processes and in the distribution of internal power.[18]

While the Senate has followed the House pattern in part, there have been far fewer changes in recent years. The Senate differs significantly from the House in that senators tend toward greater autonomy and individualism, under the umbrella of electoral politics and its associated forces, such as the emergence of a two-party system in the South. Because of its smaller size and collegial style, the Senate can better tolerate personal idiosyncracies and operate in an orderly manner without strong leaders — a situation less likely in the House. Also, because rules are much less controlling in the Senate than in the House, the Senate can effect change in more diverse, less visible ways than the House. Indeed, the Senate was *designed* institutionally to play a different role from the House: as a body, it serves as a check upon the "majority rules" flavor of the House. Its rules and customs virtually require compromises and accommodation to minority views and interests. Smaller size (100 senators) and longer tenure not only allow but encourage more informality than in the House. Senators get to know each other personally to a much greater extent than do representatives.

Overall, it can be said that the reforms of the 1970s changed Congress, particularly the House, in some fundamental ways. In the House, the coexistence of subcommittee government and strong central leaders has created the potential for a dynamic fluidity in power arrangements. This seems to be a period of divided power within Congress, where large swings in power can take place in response to external forces. Such swings, conclude Dodd and Oppenheimer, probably are most evident in the House. The Senate is better insulated from outside forces, but it nevertheless follows the House pattern to some extent.[19] Despite the six-year term, senators now spend far more time campaigning than in earlier decades; this reflects that higher Senate turnover rates than in the House in recent years. Finally, junior senators have been much less inclined to take minor roles and to wait patiently for seniority.

THE IMPACT OF DIVIDED GOVERNMENT ON CONGRESS

Among the external forces alluded to above is the extended period of divided government that the nation has experienced. Between 1980 and 1992, during the Reagan and Bush administrations, Republicans dominated the presidency while Democrats controlled Congress, the longest such period in modern times. The result of this partisan division between Congress and the President has been an extended period of policy stalemate in a number of areas. The election of 1992 brought this period to an end, although its legacy of problems is likely to shape the agenda of the incoming Clinton administration.

During each year in the 1980s and early 1990s the nation's budget deficit increased. Yet the 1988 and 1990 elections produced no political outcome that could lead to breaking the policy stalemate and addressing the budget problem in fundamental ways. The election of a Democratic President together with Democratic majorities in both the House and Senate in 1992 may end this stalemate, although the deep divisions in the electorate on which it was based are far from healed. The irony of this situation is that it would be difficult to find a member of Congress or senior member of the executive branch who does not believe the budget deficit is one of the central governmental problems of our time. Yet through the late 1980s and early 1990s, our country has been caught in a massive intragovernmental gridlock largely put in place by the electorate.

During this period, national policymaking in virtually every area has been held hostage by the massive budget deficit "overhang." The ability of Congress to serve as an incubator for legislative proposals and policy initiatives that involve spending money has been greatly constrained. This ties in with the earlier noted trend toward centralization and toward a focus on those committees that deal with budgets and appropriations and has been styled the "fiscalization" of national policymaking.[20]

CHANGES IN MEMBERS

At least some of the changes just described, including the dynamic fluidity of Congress, are due to a large infusion since the early 1970s of younger legislators in both the House and Senate, legislators who are dedicated to reform and who take a more activist approach to being a member of Congress. Notwithstanding the strident calls for reform from term-limit advocates (at least some of whom are little more than "sore losers"), over the past 10 years or so Congress has changed considerably in certain crucial respects. There are many new faces: nearly half of all members in 1992 took office since 1981; most committee chairs and nearly all party leaders came to their positions after 1981.[21] This is not an institution totally dominated, as it was at times in the past, by aging elders and long-time office holders.

Both the Senate and the House are more egalitarian institutions than they used to be. Relatively junior members in both bodies now play important roles, often in an entrepreneurial mode comparable to that of their private sector counterparts in large and small firms. These trends have opened up the political system to more participants and attracted more talented individuals, and less public business is conducted behind "closed doors." In short, Congress has become more democratic, more accountable, and more open.

Yet these trends also have their negative overtones and serious costs. Former Congressman and now New York University President John Brademas, for example, has observed that "Congress is more and more becoming a place of independent contractors, each Member intent on constructing his record in a manner most pleasing to the eyes of his constituents or special interests but without regard to his responsibility to serve the national well-being." Brademas worries that if members of Congress pursue their individual goals and are unwilling or unable to follow leadership in support of common positions, "we will never be able to construct coherent policies to deal with genuine national problems." Another practical cost noted by Brademas and others is the greatly increased "time and effort required to get things done, not a trivial consideration in a period when the sheer volume of problems that government must address threatens to grow beyond manageability."[22]

At the insistence of younger reformers in both houses, power, in the form of subcommittee chairs (about 150 in the House and 100 or so in the Senate),[23] has been distributed more widely. Among the consequences of this power redistribution — and other reforms noted earlier — has been a more dispersed, more fragmented, and more open institution. One implication of these changes is that it is easier for outside interest groups and advocates to introduce their ideas into the legislative process and have them acted upon.

CONGRESS IN POLICYMAKING

Analysts of Congress and the presidency have devoted much time and attention to the roles and relative influence of the two branches in their exercise of political power. Walter J. Oleszek has said, "Much has been written about the growth of executive power in the twentieth century and the diminished role of Congress, but in fact there has been a dynamic, not static, pattern of activity between the legislative and executive branches. First one and then the other may be perceived as the predominant branch." But Oleszek and others have argued persuasively that at any given time, such appraisals underestimate the other branch's strategic importance.[24] The American political system is best considered largely as a combination of congressional and presidential government. Yet Congress is elected separately from the President and has constitutionally based powers, and it is important to understand that Congress has a substantial degree of independent policymaking authority. Historically, this has been particularly true in a number of areas related to science and technology (e.g., environment, energy, health, research and development organization, science and mathematics education, and economic competitiveness).

Over the past two centuries, the House and Senate have used their constitutional mandate to devise their own sets of rules, procedures, and conventions — formal and informal — in order to operate. As Oleszek has noted, "these rules and conventions establish the procedural context for both collective and individual policymaking actions and behavior." More specifically, rules and procedures serve many purposes in Congress: to provide stability, legitimize decisions, divide responsibilities, reduce conflict, and distribute power.[25] In order to work effectively with Congress, it is important to be aware of the importance of rules and procedures and to acquire at least a passing acquaintance with those that may be relevant to your concerns. A splendid source for gaining such familiarity is Oleszek's *Congressional Procedures and the Policy Process* (see note 1.).

Under changes in the rules and procedures of recent years, Congress has dispersed power to many members and has opened itself to greater public observation and inspection. In today's Congress, many members, junior and senior alike, can exercise initiative and creativity in both lawmaking and legislative oversight activities. Overall, this means that the representative or lawmaking role of members has been enhanced under decentralization: members can easily obtain diverse and competing views, independently of the communication channels managed by the leadership. But decentralization and openness carry the usual price of more democracy: it frequently takes Congress longer to formulate public policies, compromises with the President are more difficult to achieve, and policy stalemate can result.[26]

In summary, Congress, in its policymaking role, displays the features, traits, and biases of its membership, rules and procedures, structure, and constitutional underpinnings. It is bicameral with divergent electoral and procedural traditions. Congress is representative, with respect to both issue constituencies and geographical interests — especially the latter. It is decentralized, power is dispersed, and there are few mechanisms for integrating and coordinating its policy decisions — especially as they cut across traditional issue and organizational boundaries, within and outside Congress. Finally, Congress is often reactive, reflecting conventional or elite perceptions of issues and problems.[27]

One negative aspect of this is that congressional policymaking often can be overly driven by localism; afflicted with partial, piecemeal solutions when the problem calls for an overarching, national solution; saddled with ceremonial, "make people feel good" solutions which solve nothing; and frozen in a state of near paralysis pending the arrival of a crisis. However, history suggests that when the crisis finally arrives, Congress can respond quickly and, most of the time, appropriately. Moreover, Congress provides the place and the circumstances to hammer out consensus on controversial issues — taking months, years, or decades, if necessary.

While this "hammering out" process is taking place, Congress may appear to be in disarray and paralyzed. The public and the media may be clamoring for "action" that Congress cannot produce. But the fact is that Congress is often doing a good job of mirroring, i.e., representing and reflecting the fact that no consensus exists on an issue, or set of issues, among the American people. This mirroring is directly related to the political gridlock issue discussed earlier.

In the meantime, while consensus and compromise are being sought, Congress serves as ombudsman for the nation's citizens in dealing with their national government. It offers the most open access of any part of the national government: there is an opportunity for individuals to take their concerns and their problems to the very top of our political system.

MAKING THE SYSTEM WORK

While the Constitution provides a marvelously constructed political system and a set of time-tested basic principles, it is itself flexible, adaptable, stretchable, and fuzzy in key places. These features also characterize the political system it established. As a result, the power relationships among the three branches of government are in a constant state of tension and dynamic flux, requiring

■ a tolerance for differences;

■ a central tendency to compromise;

■ acknowledgement that times and conditions change;

■ a broad correspondence with the popular will; and

■ a deep understanding of limits (e.g., how far to push one's views).

Over time, the balance of power among the three branches has changed — sometimes dramatically — for a variety of reasons: national emergencies, correction of abuses of power, the changing role of the nation on the world scene, and the vast growth of our economic and technological enterprises. Above and beyond the formal processes of government, informal processes and communications are necessary to make the system work. In seeking to make the system work, there is a unifying theme to the purposes of Congress and the presidency: ascertaining and implementing the people's will. Even the Supreme Court is not immune to this broad purpose. As Walter J. Oleszek has said, "The Constitution, in short, creates a system not of separate institutions performing separate functions but of separate institutions sharing functions. The overlap of powers is fundamental to national decision making."[28]

NOTES

1 Walter J. Oleszek, *Congressional Procedures and the Policy Process, 3rd ed.* (Washington, DC: CQ Press, 1989), p. 2. The discussion of basic constitutional principles closely follows Oleszek's treatment of them.

2 This principle is the major reason Congress is not subject to regulation by agencies in the executive branch.

3 See David R. Mayhew, *Congress: The Electoral Connection* (New Haven: Yale University Press, 1974).

4 Roger H. Davidson and Walter J. Oleszek, *Congress and Its Members*, 3d ed. (Washington, DC: CQ Press, 1990), p. 8.

5 *Ibid.*, p. 1.

6 Giovanni Sartori, "Representational Systems," in *The International Encyclopedia of the Social Sciences*, vol. 13 (New York: The Macmillan Company & The Free Press, 1968), p. 468.

7 Davidson and Oleszek, pp. 4-5.

8 Congress is composed of 100 senators, 435 representatives, and a total of 5 delegates from the District of Columbia, the Commonwealth of Puerto Rico, and several territories.

9 Davidson and Oleszek, p. 5.

10 *Ibid.*, p. 10.

11 Thomas J. O'Donnell, "Controlling Legislative Time," in *The House at Work*, eds., Joseph Cooper and G. Calvin Mackenzie (Austin, TX: University of Texas Press, 1981), pp. 127-30.

12 *Ibid.*, p. 6.

13 Richard F. Fenno, Jr., *Home Style: House Members in Their Districts* (Boston: Little, Brown and Co., 1978), p. 104.

14 Davidson and Oleszek, p. 7.

15 *Ibid.*

16 David E. Price in Lawrence C. Dodd and Bruce I. Oppenheimer, eds., *Congress Reconsidered*, 4th ed. (Washington, DC: CQ Press, 1989), p. 421, 436.

17 Lawrence C. Dodd and Bruce I. Oppenheimer, eds., *Congress Reconsidered*, 4th ed. (Washington, DC: CQ Press, 1989), p. xi.

18 *Ibid.*, p. 446.

19 *Ibid.*, p. 448.

20 C. Allen Schick, *Reconciliation and the Congressional Budget Process* (Washington, DC: American Enterprise Institute, 1981), p.35.

21 Davidson and Oleszek, p. xv.

22 John Brademas, *Washington, D.C. to Washington Square* (New York: Weidenfeld and Nicolson, 1986), pp. 160-61.

23 Norman J. Ornstein, Robert L. Peabody and David W. Rohde "Change in the Senate: Toward the 1990s" in Dodd and Oppenheimer, p. 20.

24 Oleszek, pp. 4-5.

25 *Ibid.*, p. 5.

26 Davidson and Oleszek, p. 305.

27 *Ibid.*, pp. 374-77.

28 Oleszek, p. 2.

CHAPTER 3

CONGRESS AT WORK

While Chapter 2 provided an overview of Congress as an institution, including its constitutional foundations, this chapter is devoted to the work of Congress and how it is done. In keeping with the theme of the "two Congresses" described in Chapter 2, the work of Congress can be divided into two broad categories: first, that of lawmaking, which includes budgets, appropriations, authorizations, other legislation, and oversight (i.e., national policy and money); and second, that of representation, or serving political constituencies (i.e., providing the kind of services that are expected of a public servant, as well as building and reinforcing political support). In discussing these two types of work, this chapter examines the growth in congressional workload, the organization of Congress and its staffing, the committee system, and the sources from which Congress obtains information and advice. An important role of the Senate that is not discussed in this guide is the "advice and consent" function for presidential appointments and international treaties.

THE MEMBERS OF CONGRESS

PERSPECTIVE

According to the popular media — electronic and print alike — members of Congress are an incompetent collection of politicians who are totally out of touch with the people they are supposed to represent. Corruption and mismanagement are words often used in reporting on Congress.

Yes, there are some incompetent members of Congress, and there have been some major institutional failures of judgment, such as not addressing problems in a timely fashion. Yes, Congress has been slow to acknowledge the need for a number of changes in its operating style, such as getting rid of patronage appointments and bringing in professional management for areas having nothing to do with the central purposes of Congress.

But few responsible media people seem willing to say, as part of balanced reporting and analysis, that individual members of Congress are, for the most part, intelligent, competent, hard working, and dedicated to serving their constituents' interests. Unfortunately, such information is simply not news and stands little chance in competing for air time or print space with congressional problems.

For years, the media has virtually ignored one of the nation's most important political stories: the fact that the political gridlock in Washington is not due to stupidity and obstinacy on the part of Congress and the presidency, but is the result of the American electorate having chosen divided government. This in turn is due to our inability, as a people, to reach a fundamental consensus on how to proceed in solving numerous national and global problems.

Too many different parts of the nation and too many different powerful political and issue groups literally have dug in their heels, refused to consider the path of accommodation, and not tolerated a view different from their own — on issues from deficit reduction to reproductive rights. Congress is reflecting or mirroring, perhaps all too well, this lack of consensus among the electorate. And if Congress appears to be in disarray, it is because the electorate itself is in disarray. This point is repeatedly emphasized because it is not generally understood.

So who are these people who are held up to ridicule by the media and the talk-show hosts and yet, as described in Chapter

2, are usually reelected by their constituents? What kinds of backgrounds do they have?

BASIC STATISTICS[1]

The average age of senators and representatives in the 102d Congress is 54, reflecting the professional status of individuals who are well along in their careers. This figure is slightly higher than in recent Congresses but is still several years lower than it was in the late 1960s and early 1970s. Since then, the number of members under 40 has increased significantly, rising from 44 in 1971 to nearly 70 in 1992; during the same period the number of members over 60 has declined from 148 to 112.

Roman Catholics (142) comprise the largest religious affiliation, followed by Methodist (75), Episcopalians (59), Baptists (59), Presbyterians (51), and Jews (39). All others add up to 115, for a total of 540: 435 representatives, 100 senators, and 5 delegates. There are two women in the Senate (one Democrat and one Republican) in the 102d Congress and 28 in the House (19 Democrats and 9 Republicans). The 1992 election increased overall minority representation in Congress by 50 percent (to 59 — including the first Native American senator) and nearly doubled the ranks of women. Six women were elected to the Senate (including the first black). Future elections are likely to continue these trends.

As in past Congresses, nearly one-half (244) of the membership of the 102d Congress describe themselves as lawyers. The next largest group, with about 190, are those who describe themselves as being in business or banking. In smaller and varying numbers, some 30 other professional categories are represented. Among the larger-sized professional affiliations are educators, media professionals (journalist, broadcaster, executive), and those who list some form of public service or public office as their profession.

Just under one-half of the members of the House in the 102d Congress (244) have taken their offices since January 5, 1981, the beginning of the 97th Congress. In the Senate, the comparable number is just under one-half (49). In other words, about half the seats in Congress have turned over during the past decade. Such data do not support charges that Congress is the bastion of entrenched, aging politicians who are out of touch with the people. It must be acknowledged, however, that such aggregate data do not reveal the great variety in election patterns throughout the country. Some seats are safe for incumbents, election after election; other seats are in a continuing "swing" status; others are close but tend to tilt toward one party; and all are subject to periodic upheaval.

THE SENATE

A brief profile of the Senate shows there are no fewer than 14 former state governors, several former mayors, and at least 5 Rhodes Scholars. Many senators had highly successful careers in a broad range of fields before coming to the Senate. The great majority have educations beyond an undergraduate university degree, and many of their advanced degrees come from the nation's premier universities.

Very few senators are professional politicians in the sense of having that they have devoted their entire careers to serving in a political office. However, about a third of the current senators have served in the House of Representatives. In age, senators range from their forties to ninety, with the average age being 57. About a third of all senators are millionaires, and some are exceptionally wealthy. While some of the wealth is inherited, much of it is self-made.

About 60 percent of all senators in 1992 are lawyers. Other professions represented in the Senate include business executives; farmers and ranchers; journalists, broadcasters, and communications industry executives; professors; military officers; and college presidents — to mention but a few former occupations. Several senators identify themselves as authors. There is one former social worker and one former astronaut, but there are none who call themselves scientists, engineers, or physicians.

THE HOUSE OF REPRESENTATIVES

Not surprisingly, the House does not appear to be as attractive to former governors of states as the Senate. In the House there is only one former governor and one former lieutenant governor. On the other hand, the House has a sizeable number of members — about 30 — who describe themselves as having had political careers or having been public administrators (having served in state and local governments in a variety of capacities as their principal career).

After lawyers, who comprise about 40 percent of the House membership, business executives or business owners are the next largest group — just over 60 individuals. Following the business group are those who have been involved with education; there are about 50 former teachers at the elementary, high school, and university levels. There are about 30 who identify themselves with the media and public relations.

Not surprisingly, there is a much wider range of occupational backgrounds in the House than in the Senate. As in the Senate, there are bankers and those in real estate and insurance; there are former judges; and there are farmers and ranchers. There also are former law enforcement officials (police and FBI), civic volunteers, accountants, pastors, morticians and funeral directors, urban planners, union officials, and television entertainers. Along with a handful of members from the health professions (doctor, dentist, and pharmacist), there are about six who identify themselves as engineers. None calls himself or herself a scientist.

GROWTH IN CONGRESSIONAL WORKLOAD

Many observers have reported on the substantial growth of the workload of Congress in recent decades. For example, Davidson and Oleszek note that this workload, which was "once limited in scope, small in volume, and simple in content . . . has grown to huge proportions."[2] This change is in direct response to the changed character of Congress.

Until the 1950s, Congress was largely a part-time institution. Members were poorly paid, and staffs were small. For the first several decades of the twentieth century, Congress stayed in session only 9 months of each 24; the members spent the rest of the time in their districts or tending to their private affairs. In recent years, Congress has been in session nearly continuously except for periodic district work periods. The average senator and representative works at least an 11-hour day while Congress is in session and often even longer in his or her state or district. Congressional staff members have comparable workdays. It once was expected that members and staff should have outside jobs; this, of course, is now prohibited, and indeed, time constraints make it impossible.

A variety of indicators may be used to measure congressional workload and its growth. Some of these include time in session, committee meetings, and floor votes. By such measures, the congressional workload has nearly doubled over the past 30 years. Committee hearings have so proliferated that members have difficulty attending the hearings of the committees and subcommittees to which they belong because of conflicting schedules. Similarly, committees continually experience problems in finding hearing rooms. Some committees have resorted to early morning (8 a.m.) and afternoon hearings in order to accommodate the many demands of their own subcommittees and those of other committees — something unheard of as recently as the 1960s.

There have been downturns in the 1980s in some indicators such as the number of bills enacted into law and the number of pages in enacted bills (see Figure 3-1), but these are a reflection not of a decreased workload but rather of its changed nature. There has been a shift to more "mega-bills" — particularly in the budget area — and

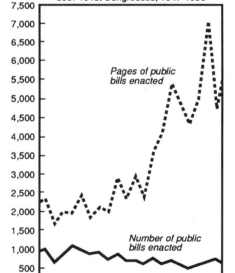

Figure 3-1. Public Bills in the Congressional Workload, 80th-101st Congresses, 1947-1990

Source: Norman J. Ornstein, Thomas E. Mann, and Michael J. Malbin, *Vital Statistics on Congress, 1991–1992.* (Washington, DC: Congressional Quarterly Inc., 1992), Figure 6-3, p. 157.

an increasing emphasis on oversight of the executive branch and on investigations that often lead to a report but not to legislation. Apart from numerical indicators, overall legislative business has grown in scope and complexity as well as in volume. The Congress of the 1990s deals with many issues that once were left to state or local government or that were not considered to be within the purview of government. Perhaps even more important, the issues of the 1990s are far more complex than those of earlier decades; more and more, they involve complicated interplay among a variety of factors — economic, political, social, and technical.

Another important dimension of the changed nature and consequent growth of congressional workload has arisen from the increased focus on constituent service and the related growth of the federal government's role in our daily lives. This increased focus is reported in academic studies of Congress, by members and staff themselves, and in such indicators as the amount of mail received and responded to by Congress. Records show that the House Post Office handled more than 200 million pieces of incoming mail in 1988 — four times the 1970 level. The comparable number for the Senate was 50 million. Davidson and Oleszek observe that "not only are constituents more numerous than ever before; they are better educated and served by faster communication and transportation. Public opinion surveys show that voters expect legislators to 'bring home the bacon' in terms of federal services and to communicate frequently with the home folks."[3]

Anyone seeking to influence or communicate with Congress must recognize that while communication lines are open, many people are using them. Those who want to get a member's attention must expect a lot of competition.

ORGANIZATION OF CONGRESS

Formally, Congress does its work through individual member offices (in Washington and in the district or state), through committees, and on the floor of the House and Senate. The informal side of congressional organization goes well beyond this, however, and includes party committees, party caucuses, and a variety of other informal groups, some of which call themselves caucuses, coalitions, conferences, and so on.[4] These informal groups may be based on party affiliation (Republican Conference, Democratic Caucus), issues (New England Congressional Energy Caucus), geography (Northeast-Midwest Coalition), gender (Women's Caucus), or ethnicity (Black Caucus).

THE HOUSE LEADERSHIP

Although members of Congress consider themselves constitutionally coequal with the President, practical considerations mean that some members are "more equal" than others. One such member is the Speaker of the House. Recognized in the Constitution and designated as next in line behind the Vice President to succeed the President, the role of Speaker carries with it special prestige and extensive power. Although the Constitution does not require that the Speaker be a member of the House, all have been. Also, while formally elected by the entire House, as a practical matter the Speaker is chosen by the majority party.

The speakership combines policy and partisan leadership with procedural prerogatives. For example, the Speaker holds unique powers in scheduling floor business and in recognizing members during floor sessions. Yet, in this modern Congress, the Speaker must be attuned to the membership and especially, but not exclusively, to the members of his own party.

Next in line in party hierarchy, an elected majority leader is the Speaker's principal deputy and serves as floor leader for the majority. (This role should not be confused with that of floor manager for a given bill under consideration by the House. There are usually two floor managers for each bill, one majority and one minority from the committee that has reported the bill.) Both House and party rules are silent on the duties of the majority leader. In practice, the job is defined by the Speaker.

On the other side of the aisle, an elected minority leader is the floor leader and the titular head of the minority party. The functions of this position have included promoting party unity, monitoring the progress of bills through Congress, and forging coalitions with like-minded members of the majority party. Another important role has been political leadership for the minority, i.e., seeking a return to majority status.

On the next rung of the party leadership ladders are the majority and minority "whips."[5] The principal whips have become elective posts in recent years ever since they have been seen as the path to majority or minority leader positions and the speakership. The other whip positions (deputy, assistant, regional) are appointive. Whips in each party meet regularly to discuss strategy and issues. For both parties, whips aid the top leaders in a number of ways: gathering intelligence, encouraging attendance, counting votes, and persuading colleagues. Whips in the House often stand at the floor entrances to signal arriving colleagues on how to vote (thumbs up for yea; down for nay). Periodic "whip notices" also are sent out advising members of upcoming floor agenda items and providing pertinent information.[6]

Since 1974, committee chairs have been elected to their positions by secret ballot in the Democratic Caucus; ranking minority members are selected by the Republican Conference. Before the 1974 reform, seniority had been the sole criterion for advancement to committee and subcommittee chairmanships. It is now only one of a set of factors used in filling these key positions. Subcommittee chairpersons and ranking minority members of subcommittees are chosen by secret ballot within each committee by the respective party committee caucuses.

Significant units utilized by the leadership of both parties in carrying out their functions currently include, for the Speaker, the Democratic Steering and Policy Committee and the House Rules Committee; for the minority leader, the Republican Policy Committee and Republican Research Committee.

While the House of the 1990s is much more open and decentralized than it once was, it is still an institution where the leadership derives much power from holding the levers on the use of rules and procedures. Effective use of these rules and procedures permits a determined majority to achieve its policy or procedural objectives. But this can be taken only so far: ultimately, the leadership — both majority and minority — must find ways of persuading members who represent different constituencies, values, and interests to support legislation before the House.

THE SENATE LEADERSHIP

The Senate of the 1990s is an institution suffused with individualism and independent operators, presenting dramatic challenges to the leaders, majority and minority alike, in performing their roles. Moreover, the rules and procedures of the Senate are much more flexible and much less structured than those of the House. (Examples of important Senate differences include unlimited debate and the "nongermane" rule. House debates always take place under rigid time constraints; those in the Senate never are constrained by time unless the Senate votes to put such constraints in place. In the House, there are limitations placed on debate remarks and bill amendments; they must be germane to the issue or bill at hand. In the Senate, there are far fewer constraints concerning "nongermaneness" of what a senator has to say.) Hence, Senate leaders rely much more heavily on personal skills, persuasion, and negotiation than on rules and procedures to carry out the Senate's business.

The president of the Senate, as defined in the Constitution, is the Vice President of the United States. However, except for ceremonial or unusual occasions, he seldom presides, and he votes only to break a tie. The Constitution also provides for a president *pro tempore* to preside in the Vice President's absence. In practice, this role has been performed by the majority party senator with the longest continuous service, although other senators serve on a rotating basis. When a senator on the Senate floor says, "Mr. President," the reference is to the president *pro tempore*. For practical purposes, the actual leadership is provided by the majority leader acting as the head of the majority party, as its leader on the floor, and as the leader of the Senate.

In the same way, the minority leader heads the Senate's minority party and acts as its leader on the floor. Neither of these positions is mentioned in the Constitution; both are subject to elections with secret ballots at the beginning of each Congress.

As in the House, there are majority and minority whip systems — although they are smaller in the Senate, as would be expected. Their purposes include gathering votes and planning party strategy. Finally, as in the House, there are Senate party caucuses, committees, and informal groups. The undergirding organizational structures in the Senate are the Democratic Conference and the Republican Conference. Also important are the Democratic Steering Committee, the Democratic Policy Committee, and the GOP Policy Committee.[7]

COMMITTEES IN CONGRESS

In the work of Congress, committees are at the center of things institutionally: in policymaking, budgets, revenues, investigations, oversight of federal agencies, and public education. While floor actions often refine the legislative products of committees, the committees are the means by which Congress sifts through thousands of bills and tens of thousands of nominations annually, along with considering issues and proposals by the hundreds.[8] It has been said or thought by so many so often that "on the Floor is Congress for show, while in committee is Congress at work."

In a comprehensive study, *Committees in Congress*, Smith and Deering found that "committees still matter" despite widespread individualism and unstructured processes in the Senate and declining specialization of members in the House, as well as diminished autonomy on the part of committees in both bodies. Members say that committees remain central to their personal goals and they continue to judge carefully the value of particular committee assignments. For example, service on an appropriations committee is a powerful attraction and appointments are vigorously contested. Also most legislative activities of members revolve around their committee and subcommittee assignments.[9] Key aspects of the committee assignment systems in both House and Senate are described below.

Committees come in a variety of types, shapes, and sizes and vary considerably in importance and influence. The basic types are standing, select, joint, and conference. The principal focus of this guide is on the standing committees, although it is also important to understand the vital role in the legislative process played by the temporary conference committees. Overall, Congress has a rather complicated organizational structure: more than 50 committees with over 250 subcommittees, each vying for its place in the sun. This total does not include all the party committees and informal groups described earlier.

STANDING COMMITTEES

Standing committees are permanent congressional entities, established by law or by House or Senate rules. Such committees continue from one Congress to the next and process the vast majority of the daily and annual business related to lawmaking, investigations, and oversight. From the thousands of bills introduced, committees choose a limited number to consider and send to the floor for possible enactment into law. Thus, they are also the burial grounds for most legislation. As of the 102d Congress, the House had 22 standing committees with 148 subcommittees. The Senate had 16 standing committees and 87 subcommittees.

The standing committees can be divided into five general categories dealing with budgets, appropriations, authorizations, revenue and taxes, and procedural/administrative matters. It is beyond the scope of this guide to describe fully the complicated interrelationships among these categories. In simplified terms: (1) budget categories and overall targets for expenditures and revenues are established by the Senate and House budget committees; (2) in varying degrees, the authorizing committees enact laws providing legislative authority for the programs and agencies; and (3) the House and Senate appropriations committees (with 13 subcommittees each) work out the details of appropriations for agency programs within these allocations and authorizations. The Senate Finance Committee and the House Ways and Means Committee are powerful and in a category of their own, as they set tax and revenue policies as well as oversee most entitlement programs and all tax-related policy incentives. One other group of standing committees deals with internal congressional administration, operational rules and procedures, and administrative and judicial procedures and organization.

SELECT COMMITTEES

Select (or special) committees are supposed to last no longer than one Congress (a two-year period — e.g., 1991–1992 for the 102d Congress). However, some take on the nature of standing committees and "just keep on going." Examples of the latter are the House and Senate committees on aging, established in recognition of the "graying of America." In the 102d Congress, the House had five select committees with a total of 11 subcommittees. The Senate had five; none had subcommittees.

JOINT COMMITTEES

Joint committees — of which there are four — have as members both senators and representatives. The chairmanship rotates between House and Senate. Joint committees are used when the House and Senate agree that the institutional interest of Congress as a whole should take precedence over the interests of either house. For the outside world the most important current joint committees are the Joint Economic Committee (the only one with subcommittees — it has eight) and the Joint Committee on Taxation; others are concerned with the Library of Congress and the Government Printing Office. It is noteworthy that while Congress has agreed upon using a joint economic committee, it has not chosen to establish a joint budget committee.

CONFERENCE COMMITTEES

Conference committees are temporary arrangements used to reconcile the differences between measures passed by the Senate and House. Appointment to conference committees is a matter of vital importance. This is where final agreements are reached about House and Senate differences. Such agreements are necessary, since before a bill can be sent to the President for signature, it must be passed by each body in identical form.

AUTHORIZATION AND APPROPRIATION COMMITTEES

While the work of the budget committees, the Senate Finance Committee and the House Ways and Means Committee have far-reaching impacts on budget, tax, and economic policy, users of this guide are likely to have more interactions with authorizing committees. And while comparable interactions with the appropriations subcommittees are not as likely to occur, there is still a strong need to follow their activities closely.

For those who would work with Congress, one of the most important distinctions to bear in mind is the fundamental difference between authorizing and appropriations committees. In principle, the authorizing committees produce legislation in the form of authorization bills that are intended to set policy, establish federal agencies and programs, and recommend budgets at certain levels. Thus a program can be "put on the books" by an authorizing committee; however, if funds are required, the program cannot proceed until a valid appropriation is enacted through the appropriations committees.

It is not unheard of for Congress to pass an authorization bill that is signed into law by the President but for which no appropriation is ever enacted. In this sense a program is "authorized" but does not really exist because the appropriations committees never voted the money to implement it. Indeed, there are billions of dollars of such unfunded authorizations "on the books." In part, this situation arises because there simply is not enough money to fund all enacted authorizations; but it also arises in part because appropriations committees often disagree with authorizing committees and, in some areas, are at times indifferent to whether or not an authorization bill has been passed; they simply proceed with their appropriation actions. Examples in recent years include National Science Foundation authorizations and certain energy bills.

COMMITTEE ASSIGNMENTS

Committee assignments are central to the organization of Congress and to the ability of members to influence policy in areas in which they are interested. Two key facts are pertinent. First, such assignments are made under party rules and processes (although in the Senate, Rule 25, which covers in detail the process of appointing senators to standing committees, is followed both by the majority and the minority). Second, committee assignments for committees and subcommittees in the House are more narrowly allocated than in the Senate. In other words, senators usually divide their time and attention among a larger number of assignments than representatives. Moreover, senators have greater latitude than representatives to influence the agendas of committees other than those on which they serve.

In the House, the majority Democrats have designated committees as exclusive, major, and nonmajor. There are three exclusive committees (Appropriations, Ways and Means, and Rules) and a member serves on such a committee as his or her sole assignment. Members not on an exclusive committee may serve on one major and one nonmajor committee — although there are various kinds of exceptions. Among the major committees are Agriculture, Armed Services, Foreign Affairs, Energy and Commerce, and Education and Labor. In the nonmajor category are Interior; Science, Space, and Technology; Budget; and Government Operations.

The Republican minority in the House uses a somewhat different approach to grouping the committees: red, white, and blue. Red committees are comparable but not identical to the majority's exclusive committees (e.g., Energy and Commerce is a red committee but is not exclusive for the majority). The blue and white committees correspond roughly to the majority's nonmajor committees. A minority member can serve on only one red committee, but on two or three blue and white committees. (Red denotes the more important committees; the others are considered generally less so.)

For both the majority and minority in the Senate, there are "A", "B", and "C" committees, as called for in senate Rule 25.

A senator is limited to two "A" assignments and one "B" assignment — but may be given a "C" assignment as well. Moreover, there are exceptions to the stated limitations. "A" committees include Agriculture; Appropriations; Armed Services; Commerce, Science, and Transportation; Energy and Natural Resources; Government Affairs; Foreign Relations; and Labor and Human Resources. Among the "B" committees are Budget, Rules, Intelligence, Aging, and the Joint Economic Committee. "C" committees are Indian Affairs and Ethics. There are no overall limits on the number of subcommittees on which senators may serve, but this is governed by the rules of each committee.

GROWING TENSION

Over the years, much tension has developed in Congress among the various sets of committees regarding their respective roles and relationships. Exacerbating the situation has been the committees' increasing interrelatedness as new systems and processes — particularly those related to budgetary, tax, and other revenue matters — have evolved during the past two decades. Further adding to the inherent tension of the congressional environment has been the way these new systems and processes were established. In many instances, they were superimposed upon existing systems; the result has been a very complicated set of processes that are difficult to understand — even for many of those who work in Congress, let alone those on the outside. Such a set of circumstances leads easily to "power plays" and competition among various members and committees.

STAFFING FOR CONGRESS

One of the most important, yet least examined, aspects of Congress is staffing. As Fox and Hammond note, in one of the relatively few studies of this subject, "Congressional staffs, both committee and personal, are an increasingly vital resource to Members of Congress."[10] As Congress has changed, staffs have expanded and changed — in distribution and qualifications. The substantial growth of congressional staff over the years is shown in Figure 3-2.

Generally, congressional staff members today are better educated, are more professional, and possess a greater variety of skills and backgrounds than staff of earlier decades. There are far fewer "political hacks" who come into the system solely on the basis of their political connections regardless of competence. Many people in key positions on the Hill have been there for years, but many others look at a congressional staff assignment as a "way station" in their career development and serve for only a few years.

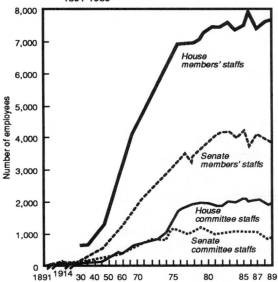

Figure 3-2. Staff of Members and of Committees in Congress, 1891-1989

Source: Norman J. Ornstein, Thomas E. Mann, and Michael J. Malbin, *Vital Statistics on Congress, 1991–1992.* (Washington, DC: Congressional Quarterly Inc., 1992), Figure 5-1, p. 127.

Members are ambivalent about the role of staff — always depending on them but sometimes resenting this dependence. There is little doubt, however, that Fox and Hammond are correct when they say, "The key aspects of what makes Congress run — activity, communication, organization, and community — in large measure involve staff. Many congressional outputs can be traced back to staff activity, where they conceptualize, write, type, and finally communicate a message."[11]

PERSONAL AND COMMITTEE STAFF

It may be useful to distinguish between personal and committee staff, but it is nearly impossible to generalize on such differences because of the wide range of organizational models and operating styles of the many hundreds of personal and committee offices. Also significant are the many differences between Senate and House styles and practices. Nevertheless, and at no small risk of oversimplifying, the following is a brief summary of distinctions between personal and committee staff:

■ Senate personal staff generally have wide latitude to play active roles in legislative initiatives. There are more barriers in the House to such activities.

■ In both the Senate and the House personal staff tend to be younger, less experienced, and more mobile than committee staff members. Also, personal staff tend to be somewhat more parochial in their outlook than committee staff. There are distinct and natural tendencies to assess the value or importance of an initiative in terms of its impact on the district or state as well as on the member's reelection prospects.

■ Personal staff who are legislative aides vary tremendously in their functions. Some are mere monitors of hearings, issues, and legislative initiatives. Others, who are perceived as clearly speaking for their bosses, can work actively and cooperatively with committee staff and other personal staff associated with their boss' committees and their boss' issues. Alliances — temporary and permanent — are used frequently to achieve desired objectives.

■ Personal staff must always keep focused on their boss' priorities, interests, and constituencies — even while working on committee business. Committee staff, while serving all committee members, must be especially attentive to "the committee agenda," the chairperson, or the ranking minority member.

■ Even as legislative aides, personal staff are necessarily more involved than committee staff with the constituent service functions of Congress.

■ Personal staff may well draft legislative amendments, committee and floor remarks for their boss, and make suggestions to committee staff on legislative strategies. However, committee agendas are generally prepared by committee staff — in consultation with the chairperson and (most of the time) the ranking minority member. In both houses, selected committee staff go to the floor. Only in the Senate may personal staff occasionally accompany a senator to the floor; this is not permitted in the House. And committee staff are mostly responsible for scheduling, planning, and organizing hearings as well as for writing reports and drafting legislation.

Constituent services are almost always done by staff (often called "case workers"), either in the Washington or district/state office. As indicated in Chapter 2, it appears that at least half a member's office workload entails constituent service. Other personal staff focus on legislatively oriented activities in working on issues and committee efforts of interest to their member. Such staff usually have some variation of the title "legislative assistant" (LA).

The staffing and structure of members' offices vary considerably, not only in overall size and types of staff, but also in the way things work. However, some commonalities are to be found. There usually is one senior staff person, more or less in charge of the staff, who is called the administrative assistant (AA). Sometimes this position is called "chief of staff"; in any event, it is important to understand that a member's AA is usually the most powerful personal staff person. Such individuals are not administrative assistants in the customary business usage, although the title is a carryover from the days when each member's staff consisted of only a secretary or two. Another common feature is the division of labor between case workers and legislative assistants. Also, increasingly, and especially in Senate offices, there is a legislative director (LD) who coordinates the work of the legislative assistants. Often the LD is also charged with helping to develop broad legislative strategies for members; he or she performs committee-related work for those members who do not have committee staff assigned to them.

Other important staff functions include press, often in the form of a press secretary, and a personal assistant to the member who handles scheduling and is often referred to as the appointments secretary. But the most important factor in congressional staffing is the member's personal work style. Some work with many staff directly, while others prefer a more hierarchical approach through the AA or LD.

LEGISLATIVE AIDES — PERSONAL AND COMMITTEE

The main focus in this guide, insofar as staff is concerned, is on legislative aides — personal and committee — rather than on case workers involved in constituent service. Overall, these staff aides can influence significantly congressional decisionmaking. Their roles are almost always influential and often critical. For example, staff has almost complete control over communications into and within a committee and personal office. Staff often have lead roles in identifying issues and developing legislative positions. Among their tasks are conducting research, gathering background data on specific legislative matters, and drafting legislation. They research and draft testimony, speeches, floor remarks, letters to constituents, and reports. In cooperation with other staff, they increasingly coordinate legislative strategy, even on key issues. They must keep track of a multitude of issues and events and be able to provide succinct briefings to their member. Finally, and perhaps most important, they are expected to offer their opinions and serve as a "sounding board" for their member or chairperson.[12] As will be discussed later, staff members are key conduits for information to and from members and key contacts for working with them.

As Figure 3-2 (through 1989) and Table 3-1 (based on 1992 congressional personnel records) show, there are about 1,500 committee staff in the Senate and about 2,200 in the House. This makes a total of about 3,700 staff who are directly concerned with the business of government — i.e., with congressional functions related to lawmaking, investigations, and oversight. While some may think this a large number, it is only about 10 percent of the number (about 40,000) often used by critics of Congress to describe a bloated staff system.

Table 3-1
Distribution of Congressional and Support Group Staff (1992)
Congressional Staff[13]

	House	Senate
Member personal staff	7,800	4,000
Committee staff	2,200	1,500
Support staff[14]	2,300	1,700
Subtotals	12,300	7,200
TOTAL (House & Senate)	**19,500**	

Support Groups[15]

Architect of the Capitol	2,500
General Accounting Office[16]	4,100
Government Printing Office[17]	5,000
Library of Congress	5,000
[Congressional Research Service	861]
[Copyright Office	527]
Office of Technology Assessment	143
Congressional Budget Office	226
TOTAL (Support Groups)	16,969
GRAND TOTAL	**36,469**

Table 3-1 also notes that House personal staff number about 7,800 or just about 18 per office, since each member receives the same personnel allowance. In the Senate, personal office staffing amounts to about 4,000 and is allocated on the basis of state population: senators from states with small populations, such as South Dakota, get a far smaller allocation than do senators from large states such as California. The overall growth in staff has leveled out since the early 1980s. The earlier growth was largely in response to growth in the role and size of the federal government and to the increasing number of executive branch agencies. In a sense, staff equals power for Congress to legislate, investigate, and oversee the operations of government.

UNELECTED LAWMAKERS?

Members of Congress simply cannot handle the heavy workload on their own; they must rely extensively on what some have called "unelected lawmakers." As Davidson and

Oleszek have observed, "their influence can be direct or indirect, substantive or procedural, visible or invisible."[18] It is particularly important that the discretionary agenda of Congress and its committees, and even that of personal offices, is greatly influenced and shaped by the staff. For example, senior committee staff members and staff directors have described in considerable detail how committee agendas are planned: the chairperson and ranking minority member will have their interests taken care of in the process, but so will the staff.

In our form of government, policy proposals arise from many sources — inside and outside the government — and congressional staff members are positioned so as to be able to advance, hinder or even stop these proposals. Staff actively engage in negotiating with members, lobbyists, outside interest groups, and executive branch officials on a wide range of issues, legislative language, report language, and political strategy.[19] In short, throughout the many stages of congressional policymaking, staff members play an active part. They are deeply involved in preparing committee agendas (legislative and oversight), planning hearings, recruiting witnesses, drafting reports and legislation, participating in drafting amendments for committee mark-up sessions, and even (in limited numbers) accompanying members to the floor when their legislation is being considered.

Parallels have been drawn between the courtiers of kings, assistants to presidents and generals, and the staffs of members of Congress. In contrast to many other organizational settings, the staff organization pyramid in congressional offices is often flat, and the principal figure has direct access to many of the professional staff. Organizational norms such as promoting on the basis of merit, seeking formal professional recognition, following standard career patterns, and selecting based on rigorous procedures are not characteristic of the congressional setting. Rather, such norms as personal loyalty, persistence, deference, special rules of courtesy, maintenance of a low public profile, and less concern with formal assigned duties and tasks seem to be associated with both committee and personal staffs. There is very much the sense of a "personal team," whether it be a personal or a committee office.

PROFESSIONALISM AND PARTISANSHIP

As staff size has increased, as staff specialization has developed, as and more professional experts have joined the congressional staff system, there is no doubt that tension has arisen between the demands of professionalism and the exigencies of a personalized and often partisan working environment. Many staff are necessarily partisan both in their orientation and in their congressional activities. Since most staff are appointed by a partisan, they may be expected to, and most often do, reflect the partisan orientation of their patron. Professional experts coming into the congressional staff system or working with it must be aware of and sensitive to these features of Congress.[20]

HEARINGS

The budget committees, appropriations committees, and authorizing committees — as well as joint, special, and select committees or their subcommittees — all hold hearings related to legislation being considered. The character of the hearings, and the range of testimony that is heard, varies widely, however.

BUDGETS AND APPROPRIATIONS

Opportunities for the public to participate in the work of budget committees and appropriations committees and their associated hearings, are limited. Agency representatives appear in order to explain and defend their programs and the President's budget request. However, outside or public witnesses presenting expert testimony are rare. Money-related congressional activities are very much an inside game.

To the extent that public witnesses are involved, it is most often through "public days" set aside by the committees. The budget committees do extend invitations to some witnesses, but the prevailing practice in the appropriations areas is to require individuals and organizations to request time, typically for only a few minutes, and generally as a supplicant rather than as an advisor. In short, there is only a very limited requirement in these types of congressional hearings for expert information and advice. When it is desired, the committees tend to acquire information informally and through special reports or investigations conducted by investigators on loan from various agencies.

LEGISLATIVE PROPOSALS

There is a substantial requirement for public testimony involving expert advice and information in congressional hearings before the authorizing committees. The more than 200 subcommittees in the Senate and House that address legislative matters have created a large and growing demand for information, analyses, and expert advice on virtually every important national and international issue — as well as for many of limited import or consequence. Accentuating this demand have been several key features of our political system noted earlier. First, one of the results of the congressional reforms of the mid-1970s was the distribution and decentralization of power among subcommittee chairpersons and substantially increased staffs. A second factor has been the extended period of divided government, with a Republican executive branch and a Democratic Congress. In a very real way, this has led to higher levels of policy conflict — as Congress seeks information and advice independently — than if a single party controlled both Congress and the presidency.

OVERSIGHT

At least equal to, if not larger than, the legislative demand is the very major requirement for expert advice, analyses, and information in the form of public testimony for oversight activities performed by Congress. These oversight functions were formalized under the Legislative Reorganization Act of 1946, which calls for Congress to perform "continuous watchfulness" over the agencies under the jurisdiction of various committees. For practical purposes, oversight can be divided into four major categories:

■ Committees involved with authorizing legislation are required to review the programs and operations of federal agencies within their jurisdiction and to recommend appropriate corrective action for identified problems.

- Financial or fiscal oversight responsibilities are undertaken by the appropriation subcommittees in the House and Senate. Similarly, for revenue-related activities, the Senate Finance Committee and the House Ways and Means Committee undertake oversight activities.

- A broad range of investigative responsibilities are assigned explicitly to the Senate Governmental Affairs and House Government Operations Committees. Their mandates range across the government and are not confined to any particular agency or set of issues.[21]

- A significant trend in recent years has been for the authorizing committees, particularly, to perform more wide-ranging investigations through special oversight and investigation subcommittees. In key areas, such subcommittees have been chaired by the chairperson of the full committee; a notable recent example is the House Energy and Commerce Committee.

In recent times, congressional committees have used their investigations and oversight powers increasingly in a large number of areas. Historically, Senator Harry Truman's investigation of procurement fraud during World War II was clearly a major stepping stone to the vice presidency and ultimately, the presidency. Richard Nixon used the Alger Hiss case in much the same way. Other members of Congress, over the years, have seen the possibilities inherent in combining wide-ranging oversight investigations with an active media interest in selected areas, whether it be alleged scientific fraud or charges of misuse of indirect costs at universities, perjury at the Environmental Protection Agency, or misinterpretation of global environmental data. Notwithstanding that such oversight activities in combination with extensive media coverage may well advance individual members' agendas, other members see these activities as a fully legitimate use of congressional power. Recent oversight hearings and investigations into scientific misconduct and university management of research funds are seen in this light.

INFORMATION AND ADVICE FOR CONGRESS

Information and advice are available to members of Congress from a wide range of sources. Some are internal, such as other members, staff, and congressional support agencies. Others are external, including the executive branch, associations, interest groups and lobbyists, private individuals, and the media. Recent studies suggest that members are relying more heavily on executive branch personnel, the congressional support agencies, and interest groups and less on consultants and volunteer experts.[22] Survey data collected during the preparation of this guide are consistent with these observations.

This does not mean there is no role for scientists and engineers as volunteer experts — but it is another indication of the difficulties in getting one's message across. An important aspect of this topic relates not to how Congress gets information, but rather to how it sifts, sorts, evaluates, and validates information and advice. Such matters as familiarity, track record, and trust influence receptivity. All these things have important implications for scientists and engineers, as is discussed in the next chapter.

INTERNAL SOURCES: OTHER MEMBERS AND STAFF

There is a vast, uncharted — and little studied — flow of information and advice among the members themselves. Over time, members develop their personal networks, along with assessments of who is reliable, trustworthy, and knowledgeable and who is not. They depend on these networks and contacts for everything from advice on election campaigns to what to do about a given issue. Sometimes their staff aides know about these contacts, but often members keep both their sources and the specific advice private. The House and Senate floors, the cloakrooms, and even the gymnasiums (from which staff members are excluded) provide settings for such conversations.

An important implication for outsiders seeking to work with Congress is that some members have more status than others and are more influential in certain policy areas then are other members. Members cannot have expertise in all areas, so they turn to trustworthy or influential colleagues as necessary. This status is independent of formal organizational roles and depends more on how a member is viewed on a personal level by his or her colleagues. Assessing how influential a member is in these terms is not easy but should be part of one's intelligence-gathering operation. Putting it bluntly, you reduce your chances of success if your main contact is not well thought of or is seen as having little influence within Congress. Identifying your issue too closely with a member held in low regard by colleagues may give it the kiss of death.

Although members, as noted earlier, are sometimes ambivalent about their dependence on staff, there is no question that congressional staff constitute one of the main sources of information and advice for members. Levels of influence vary widely among staff members, however, and sensitivity is required in choosing which ones to work with.

CONGRESSIONAL SUPPORT AGENCIES

The four congressional support agencies have become highly integrated into the operations of Congress.[23] Their nonpartisanship, objectivity, and responsiveness to the requests of members make them valuable resources that members hold in high esteem — although each agency has encountered tensions and even hostility from time to time. One explanation of members' overall positive appraisal for the agencies may lie in the following observation by Davidson and Oleszek:

> Unlike committee or personal aides, these agencies operate under strict rules of nonpartisanship and objectivity. Staffed with experts, they provide Congress with analytical talent matching that in executive agencies, universities, or specialized groups.[24]

It is worth noting that these agencies are not chartered to provide information or analysis to the executive branch or to the public. Even so, many of their products are widely available — formally and informally.

Congressional Research Service

The oldest and most frequently used congressional support agency is the Congressional Research Service (CRS), a unit of the Library of Congress. Established in 1914 as the Legislative Reference Service, it focused initially on the preparation of law indices and digests. Although this remains a major service of CRS, its mandate has expanded to include providing an extensive array of research services, preparing analyses, monitoring issue areas and gathering pertinent data, and preparing legislative materials. CRS is one of the two support agencies — along with the General Accounting Office — that can provide direct assistance to both members and committees, not just to committees. Further, both of these agencies can assign individuals on detail (or loan) to congressional committees for special purposes such as preparing a major report, planning a major hearing, or conducting an investigation. (It should be noted, however, that this practice on the part of CRS has greatly diminished in recent years).

The Legislative Reorganization Act of 1970, which changed the agency's name to the Congressional Research Service, also gave it a new emphasis on policy research and analysis. Since 1970, CRS's staff has grown from 200 to nearly 900, and a number of new units have been created, including a Science Policy Research Division. A sizeable number of CRS staff members come from academic backgrounds or have credible academic credentials. More and more, they operate in a "think tank" mode for Congress.

CRS offers a great deal of variety and versatility in its services. In one sense it reflects its library origins by providing quick responses to thousands of requests annually for factual information. In another way, its policy research and analysis roles are illustrated by the publication of a large number of reports — self-initiated as well as requested by members and committees. These include the widely used *Issue Briefs*, reports on many topics important to Congress, a *Digest of Public General Bills and Resolutions*, and summaries of major legislation considered by Congress. While technically CRS reports are not available to the general public, in practice, many of them are widely distributed outside Congress — often through members' offices.

The short preparation times, urgent deadlines, and confidentiality of many CRS products often preclude a formal outside review. To compensate, CRS has established a rigorous internal review process. First, a report is reviewed by peers within the author's division and, if appropriate, by analysts in other divisions for accuracy and analytical quality. After this peer review, the report is given a division-level review to ensure that it meets division standards of technical accuracy and congressional needs.

Next, the Review Section of the Office of Policy reviews the document for compliance with CRS standards of balance and objectivity as well as for responsiveness to congressional needs, clarity, and timeliness. If the review office believes that additional peer review is necessary for technical content, it has the authority to request such a review.

When time and confidentiality considerations permit — for example, for reports and issue briefs that are written in anticipation of congressional requests — outside review is strongly encouraged. Usually such review is initiated by the analyst, but division management and the review office also have the authority to request such a review. The other congressional support agencies often participate in this outside review process.

In addition to supplying written products, CRS briefs members and their staffs, analyzes pending issues, and holds numerous seminars for members and staff on a wide range of subjects. Finally, CRS works with committees in helping to establish their agendas, in suggesting areas they might want to investigate, and in providing checklists of legislation under the jurisdiction of various committees that requires revisiting or reauthorization.[25]

General Accounting Office

The General Accounting Office (GAO) was established in 1921, together with the Bureau of the Budget (now the Office of Management and Budget), under that year's Budget and Accounting Act. Directed by the Comptroller General, who is nominated by the President and confirmed by the Senate for a 15-year term, GAO, as described by Oleszek, "conducts audits of executive agencies and programs at the request of committees and Members of Congress to make sure that public funds are properly spent."[26] But perhaps more important, GAO is Congress's premier field investigator, domestically and internationally.

GAO maintains a staff of around 5,000 and issues more than 1,000 reports a year. The agency deals with a wide range of issues, focusing mainly on eliminating waste and fraud in government programs and improving program performance.[27] Subject to approval by the Comptroller General, large teams of investigators may be assigned to work on investigations extending over a number of years in close coordination with congressional committees.

GAO reviews can lead to congressional hearings, enactment of legislation, and significant administrative changes in the ways that executive branch agencies do business. Under the Legislative Reorganization Act of 1970, GAO was given an expanded role in providing oversight assistance to Congress. The Congressional Budget and Impoundment Control Act of 1974 propelled GAO even more deeply into evaluating programs, obtaining program data, and assisting in the congressional oversight function.

The products of GAO's reviews can take several different forms: testimony by GAO representatives before congressional committees and subcommittees; oral briefings for members of Congress and staff, particularly on the progress of requested review; and written reports addressed to Congress, a requester, or an agency. Written reports range from strict statements of the facts to detailed analyses of the data obtained, including conclusions and recommendations. While they may vary in format, content, and complexity, reports, and the work performed in producing them, are subjected to stringent quality review procedures within the agency. Federal agencies and other parties affected by or related to GAO reviews are often given the opportunity to comment on draft reports — especially when the issues are sensitive or controversial or include significant recommendations for action by the agency head or Congress.

An especially important aspect of all GAO products is the independent stance of the agency. The Comptroller General is appointed for a 15-year term for just this reason — independence. In seeking to provide useful and credible analyses and information to Congress, the Comptroller General and the agency insist on planning, performing, and reporting their work independently and objectively. Thus GAO maintains discretion in determining how and by whom the audit or evaluation work is to be performed. This discretion extends to deciding what is to be included in all GAO's products — oral and written.

In contrast to the somewhat limited distribution of CRS products, Congress has directed that GAO reports be given wide distribution. As is the situation throughout Congress, the workload of GAO has grown enormously during the past two decades. However, not all of this growth can be attributed to Congress. While roughly four-fifths of GAO's workload is directly for Congress, the rest is related to other purposes, including GAO serving as a major audit arm of the federal government. Also, a growing focus on "the budget as policy" and the new budget processes being used within Congress and the executive branch have led to greater interest and willingness on the part of members and staff to rely more on the sophisticated program evaluation and data-gathering techniques of GAO.[28]

The fundamental nature of GAO investigations, evaluations, and reports is quite different from those of CRS. Moreover, the special tenure and high status of the Comptroller General provides the agency with a significant degree of independence from both Congress and the executive branch. GAO maintains independent management control over even those investigations it undertakes for congressional committees and members. And whereas CRS reports do not recommend policy,

GAO reports often put forth controversial policy recommendations. It is no accident, then, that GAO and agencies of the executive branch often end up in disagreements and confrontations over findings, conclusions, and recommendations arising from GAO investigations.

Office of Technology Assessment

Considerably newer than CRS and GAO is the Office of Technology Assessment (OTA). Establishment of the office in 1972 was one of the main products of the "technology assessment" movement of the 1960s. The organization was created by the Technology Assessment Act of 1972, which was based on the premise that Congress needed to "equip itself with new and effective means for securing competent, unbiased information concerning the physical, biological, economic, social, and political effects" of the application of technology.[29]

OTA is managed in fundamentally different ways from the CRS and the GAO. At the outset, Congress placed itself in control by setting up the Technology Assessment Board (TAB), a bipartisan body made up of six representatives and six senators. Despite early concerns that OTA might become the political creature of those members on the board, it has not turned out that way.

In practice, the board has confined itself to major policy shaping — selecting the director, quality control of processes and outputs, and broad priority setting. If anything, the board has served as a political shield against attempts to use OTA as a political tool for any given point of view. Also, in contrast to the CRS and GAO, the OTA has an outside advisory committee consisting of the Comptroller General, the director of CRS, and 10 distinguished public members.

As another major contrast with its fellow support agencies, OTA's internal staffing is limited by law to about 150. It makes extensive use of individuals who come to work for the organization for limited periods of time (contractors, consultants, and fellows). Unlike GAO and CRS, OTA does not work directly in response to individual member requests; congressionally requested OTA studies are accepted only from committees and through the TAB. Also, while CRS is geared to deal with quick responses and informal questions, OTA is not.

Much of OTA's work is conducted through contract studies which are then synthesized and reviewed internally before publication as OTA reports. The study review process leading to a report is rigorous and often lengthy. At the beginning of an assessment, OTA assembles an advisory panel of 15 to 20 stakeholders and experts to ensure that its report will be objective, fair, and authoritative. The advisory panel typically meets with the project staff three times during a study. Its members help to shape studies by suggesting alternative approaches, reviewing draft documents, and critiquing draft reports. OTA panels include members with widely differing economic and political interests, disciplinary backgrounds, and institutional affiliations. Efforts also are made to include people with a range of ages, women as well as men, ethnic minorities, and people from different regions of the country.

In addition to the advisory panel, OTA often convenes workshops at which experts focus on a particular aspect of a study. OTA believes that the involvement of people with differing backgrounds and interests greatly strengthens its work. Thus, in addition to the panel and workshop members, as data is collected or contract reports are done, other stakeholders and experts are asked to review draft contractor reports, chapters of the assessment, and other documents. In addition to technical experts, OTA seeks reviewers who are expected to disagree with a given piece of work — perhaps because they have already expressed public disagreement with the author or because their background or institutional affiliation reflects a conflicting viewpoint.

As OTA studies develop, the draft chapters are reviewed externally at least once, usually several times. Between 50 and 200 stakeholders provide critiques on various aspects of a typical study. OTA staff in related subject areas also provide comments. After the external review, OTA conducts a formal internal review involving the responsible program manager, assistant director, and the director. Circulation of a courtesy draft to the other assistant directors allows anyone on the OTA staff to join in the review. After the internal review process is completed, OTA sends the final draft to the Technology Assessment Board for review before release for publication.

In recent years, under the guidance of its board and an outside advisory committee, OTA has moved far beyond its original, narrowly defined legislative mandate. It has undertaken studies of some of the more complex policy questions facing the nation and the world, especially those rooted in scientific and technical considerations. Its reports have been acclaimed — and attacked — and are distributed and quoted widely, domestically and internationally. In spite of this, however, some members of Congress and their staff misunderstand or are little aware of OTA's role and its products.

Congressional Budget Office

The Congressional Budget Office (CBO) is the newest of the four major congressional analytic support agencies. It was established under the Congressional Budget Act of 1974 and was designed to be an integral part of a new approach to dealing with budget activities in the congressional setting. At least in part, it was intended to be the congressional counterpart to the Office of Management and Budget (OMB) in the Executive Office of the President.

In comparison to the overall charter of OMB and even compared to some of the other congressional support agencies, CBO's mission is relatively narrow: to provide economic and budgetary information in support of the congressional budget and legislative processes. The subject matter of the agency's work is rather broad, however, given that the budget of the federal government covers a wide range of activities and plays a major role in the U.S. economy as well as the international economic scene.

The Congressional Budget Act was intended to strengthen the ability of Congress to deal with the federal budget and to restore the balance of budgetary power, which a number of analysts and participants believed had tipped too much toward the executive branch.[30] As part of this new framework, CBO is intended to give independent budgetary assistance, economic analysis, and policy analysis to Congress. As the budget has gained in political significance in recent years, CBO reports and analyses have received increasing attention in Congress, in the media, and in government in general.

Like OTA, and in contrast to GAO and CRS, the Congressional Budget Office does not respond to individual member and staff requests. Its work and lines of communication are with the House and Senate committees. With a staff of about 225, it publishes reports on the budget, including scorekeeping; makes estimates of tax receipts and government expenditures; makes economic forecasts and projections; makes estimates of the cost of proposed legislative proposals (not always to the liking of their sponsors); and conducts background studies and analyses in a variety of major policy areas.[31]

Like the General Accounting Office, CBO is frequently asked to testify before congressional committees. While the testimony is most often in connection with an ongoing or completed study, sometimes special analyses are prepared for such committee appearances. However, the most important CBO product is very likely its annual report to the House and Senate budget committees. One component of this report is a document providing economic and budget projections for the next five years; the other component is a plan for reducing the budget deficit.

CBO has established a detailed set of procedures related to preparation of its products and maintenance of quality standards. These include the use of an extensive set of detailed guidelines and questions to be addressed, internal and external reviews, extensive coordination among CBO units, and full clearance by the director's office before a report is released publicly.[32]

There seems to be little doubt that CBO has become a major force not only in congressional budgetary affairs but on the national scene as well. Its analyses, its "scorecards," and its written reports on important policy issues have become essential sources for those inside and outside Congress. If there is an overall tone to CBO operations, it more resembles the hard-nosed, skeptical fiscal conservatism of the OMB than it does the expansive, program-initiating orientation of some congressional authorizing committees.

EXTERNAL SOURCES

Executive Branch

Congressional committees are often organized to correspond, at least approximately, with the jurisdiction, mission and functions of one or more executive branch agencies. Examples include the agriculture, defense, interior, labor, and small business committees in both House and Senate. This correspondence can, over time, result in close working relationships among congressional committees, a member's personal office (especially a chairperson or subcommittee chairperson), the agencies overseen, and the interest groups most directly affected by the agency.

Scholars in political science and public administration have long studied these types of tripartite relationships, and they are well documented in the literature of these fields. For example, the arrangements have been characterized as the "so-called iron triangles — a shorthand term that embraced a wide variety of relationships."[33] And whatever form they took, they entailed a large number of policy arenas. The lines of communication between congressional and agency staff and between members and top agency officials are essential sources of information for both sides — even if they are at times a cause of anger in the White House, whether in Democratic or Republican hands.

For the outsider, these lines of communication mean that it is not always necessary to contact Congress directly in pursuing one's interests. It may well be preferable to work indirectly, through an agency staff member who already has the access, trust, and skill in maneuvering through the turbulent waters of Capitol Hill.

Interest Groups

Interest groups, including lobbyists, play an active part in congressional activities, and as the role of the federal government in society has grown in recent decades, the number of such groups seeking to influence Congress has mushroomed. Interest groups perform important, virtually indispensable functions: informing Congress and the public; stimulating public debates on key issues; and making available to Congress all sorts of information and analyses on proposed legislation and oversight activities.[34] Specifically, Congress can look to interest groups for draft legislation, analyses of competing proposals (especially those of their adversaries), draft speeches, answers to questions (usually provided promptly), questions to ask at hearings, and a host of other things. In short, an interest group that is "plugged in" can operate in some ways as an extension of a member's or committee's staff.

While it is true that interest groups can provide a variety of benefits to those in Congress, it is also important to understand the need for caution and care in these relationships. A key challenge for members and their staffs is to use the information and assistance provided by interest groups without becoming bound to their wishes and specialized agenda. In a direct warning, E. Douglas Arnold states that in regard to governmental resources, "interest groups usually have their own ideas about proper allocation, and they seldom coincide with Congressmen's predilections."[35]

Other troubling aspects of the participation of interest groups in our political system derive from the confluence of the increasing costs and the consequent growth in the role of money with the diminishing influence of political parties in election campaigns. Special-interest groups have increasingly stepped into the breach. As John Brademas suggests, "Such groups often concentrate on single issues about which they feel strongly, and they attempt to focus a congressional election solely on those issues."[36] This preoccupation with single issues is surely one of the most worrisome features of the U.S. political scene.

However, for better or for worse, with over 2,000 special-interest group offices in Washington, this feature of our political system is solidly in place. Of course, not all interest groups are equal in terms of access or influence. Oleszek's conclusion, widely shared by members and staff alike, is that an interest group's ability to influence congressional activities is based on the following factors:

- The quality of arguments and information;
- The size, cohesion, intensity of the organization's membership, and the ability to marshall them;
- The group's ability to develop alliances — temporary or permanent — with other organizations;
- Its financial and staff resources; and
- The vision and shrewdness of the leadership.[37]

A closely related aspect of interest groups not discussed in detail here is the rise of the PAC — political action committee. More than 4,000 of these entities have been established by various interest groups to raise and contribute funds to political campaigns. The continuing escalation in the costs of running for Congress has led many members and candidates to turn to special-interest groups, large numbers of which are more than willing to supply campaign money. The 1980s and 1990s have seen a mind-boggling growth in the number of PACs and in the amount and significance of their donations. It is ironic that this development was not foreseen when limitations were placed on individual contributions as part of the 1974 post-Watergate political reforms. John Brademas states that "to observe physical evidence of PAC power, one need only walk by a congressional committee room when legislators are writing the final version of a bill to see the host of lobbyists watching interestedly and reporting every move that Representatives and Senators make."[38] However, the PACs themselves are watched carefully by the media and various self-appointed interest groups since their activities are part of the public record.

The Media

No single influence is more important to Congress than the media. To put it simply, members and staff live, in direct and important ways, by what they read, watch, and hear in the media — as well as by what they create for the media. This involvement ranges from current newspaper headlines and the nightly television news to intensively analyzed issues in small-circulation but influential journals.

NOTES

1 Much of the statistical data cited in this section is drawn from Norman J. Ornstein, Thomas E. Mann, and Michael J. Malbin, *Vital Statistics on Congress, 1991-1992* (Washington, DC: Congressional Quarterly Inc., 1992).

2 Roger H. Davidson and Walter J. Oleszek, *Congress and Its Members*, 3d ed, (Washington, DC: CQ Press, 1990), p. 3. The summary treatment of congressional organization in this chapter follows that of Davidson and Oleszek.

3 *Ibid.*, pp. 31, 34.

4 The Constitution says nothing about political parties, but they were devised early in the history of the country as a practical matter to deal with political organization.

5 The term "whip" was adapted from usage that arose during the eighteenth century in the British Parliament — as are many other procedures, customs, and practices of the U.S. Congress. Originally used in fox-hunting, the "whip" kept the pack of hunters and dogs under control or "in line." The sense of the adaptation seems rather obvious, although a physical whip is no longer used — despite the occasional urge of a modern congressional "whip" to have such a capability.

6 Davidson and Oleszek, pp. 159-66.

7 *Ibid.*, pp. 167-76.

8 *Ibid.*, p. 195.

9 Steven S. Smith and Christopher J. Deering, *Committees in Congress* (Washington, DC: CQ Press, 1984), p. 271.

10 Harrison W. Fox, Jr. and Susan Webb Hammond, *Congressional Staffs: The Invisible Force in American Lawmaking* (New York: The Free Press, 1977), p. 1.

11 *Ibid.*

12 *Ibid.*, p. 2.

13 Information based on data acquired from the offices of the Clerk of the House and the Secretary of the Senate. Numbers are rounded to the nearest hundred.

14 Support staff includes police, cafeteria workers, and various types of trades such as electrical, carpentry, and plumbing.

15 Information based on data acquired from each of the support groups.

16 The GAO estimates that about 80 percent of its work force (which totals about 5,100) is connected with service for Congress. This estimate is the basis for the number 4,100.

17 The GPO staffing number is based on the assumption that 100 percent of the GPO's efforts are intended to serve Congress.

18 Davidson and Oleszek, pp. 218-19.

19 *Ibid.*

20 Fox and Hammond, pp. 157-59.

21 Walter J. Oleszek, *Congressional Procedures and the Policy Process* (Washington, DC: CQ Press, 1989), pp. 263-64.

22 Fox and Hammond, p. 122.

23 For a more in-depth discussion of these agencies, see Carnegie Commission on Science, Technology and Government, *Science, Technology and Congress: Analysis and Advice from the Congressional Support Agencies* (New York, Oct. 1991).

24 Davidson and Oleszek, p. 230.

25 Fox and Hammond, pp. 131-32.

26 Oleszek, p. 271.

27 *Ibid.*

28 Fox and Hammond, p. 133.

29 *Science, Technology, and Congress: Analysis and Advice from the Congressional Support Agencies*, p. 23.

30 *A Profile of the Congressional Budget Office* (Sept. 1990), pp. 1-9.

31 Fox and Hammond, pp. 134-35.

32 "CBO Manuscript Procedures," in *CBO Administrative Manual* (Washington, DC: Congressional Budget Office, n.d.), pp. 3-10.

33 Roger H. Davidson in *The New Congress*, Thomas E. Mann and Norman J. Ornstein eds. (Washington, DC: American Enterprise Inst., 1981), p. 105.

34 *Ibid.*, pp. 136-37.

35 E. Douglas Arnold in Mann and Ornstein, p. 280.

36 John Brademas, *Washington, D.C. to Washington Square* (New York: Weidenfeld and Nicolson, 1986), p. 157.

37 Oleszek, p. 41.

38 Brademas, pp. 158-59.

CHAPTER 4

WORKING WITH CONGRESS

This chapter is designed to help you learn how to work better with Congress in influencing its actions as they may affect you or your organization. Remember, doing so is not a privilege; it is your right. But exercising your right effectively takes skill, knowledge, and practice. Formal and informal meetings with members and staff, telephone contacts, correspondence, and contacts with state and district offices are all discussed in individual sections of this chapter. Hearings and testimony are covered separately in Chapter 5.

Each of the sections in this chapter is organized as a series of "do's" and "don't's" intended to help you make the most of your contacts with Capitol Hill. Many of these pointers, rules, and suggestions are based on the discussions of the culture and workings of Congress in Chapters 2 and 3; they are numbered for ease of reference, not because some are necessarily more important than others. In addition, the chapter draws on the experiences and opinions of members of Congress, their staffs, and skilled professionals who work with Congress, much of which came out of questionnaires and personal interviews conducted expressly for this guide. In short, this advice is based on experience, some bitter and some sweet.

One important goal to keep in mind is to become a *recipient* of congressional telephone calls and requests for information or assistance. This is reflected in what one senior staff person said: "I live by my Rolodex." This individual and many others emphasized how much they use the telephone in contrast to reading letters or reports. Members and staff operate within complex networks of information flows. Over time, each member and staff member develops a network of individuals on whom he or she comes to rely for advice, information, suggestions for prospective witnesses, evaluations of reports, assessment of people, etc. If you can become a valued source based on your performance, reliability, and credibility, then many of the entree problems discussed elsewhere in this chapter will vanish.

SEVENTEEN CARDINAL RULES FOR WORKING WITH CONGRESS

While certain principles apply to specific types of interactions with Congress — meetings, telephone conversations, and correspondence, for example — a number of more general rules apply to all these situations. All of these rules derive from one overall principle: **Always think in terms of improving the chances of acceptance of your ideas, suggestions, or proposals.** Whatever mode you use for working with or contacting those in Congress, your overriding concern should be: "How can I improve my chances for communicating my ideas successfully and getting them accepted?" In operational terms, this means you should:

> "Time is at a premium. Present your best case in a positive, honest fashion, and as quickly as possible."
>
> **SENATOR MARK HATFIELD (R-OR)**

1. **Convey that you understand something about Congress.** A recurring complaint among members of Congress and their staffs is that so many who come to see them seem to know so little about Congress. Members and staff do not expect you to be an expert on Congress, but they do appreciate (and have more respect for) those who display an awareness and understanding of what is going on — particularly with regard to the conditions members and staff face. Among other things, these

conditions include severe time constraints, competing demands for legislative and budget priorities, and the imperatives of reelection. Citing what may be an extreme case, a staff member explained why one visitor received a very negative reception: "This guy didn't realize that Representatives have to face an election every two years!"

2. Demonstrate your grasp of the fundamentals of the congressional decisionmaking system, especially the need for compromises and trade-offs. Members and staff say that one of the most difficult things to get scientists and engineers to understand is the tough reality faced by individual members in balancing competing interests, building working alliances, and achieving acceptable compromises. Among their comments are that "scientific elites don't acknowledge other legitimate interests"; "there is a lack of understanding that they are in a competition like everyone else"; and "scientists are perceived as just another constituency." Finally, as one staffer pointed out, there is "a frequent misperception that a Member will vote against one of his or her constituencies if only you will give them the correct facts." Unlike science, politics cannot be reduced to empirical facts and figures. Indeed, it is rare that an initiative is not substantially modified through compromises and trade-offs before a final policy decision is made or a law is enacted. This means that you may lose even if you have a good case and a good relationship with the member. It also means that you should not take it personally and should keep trying. Persistence can pay off.

3. Don't seek support of science as an entitlement. This may seem obvious, but it is a problem that occurs with sufficient frequency to require highlighting. Even if the word "entitlement" is not used directly, members and staff react negatively when they are presented with arguments in support of science that they see as being cast in "entitlement terms." In their words, "support for science should be presented in terms of helping to meet national needs, or to achieve societal goals, not as an entitlement owed to scientists," and scientists and engineers should not "convey an attitude of being inherently deserving in contrast to other seekers of the public largesse."

4. Don't convey negative attitudes about politics and politicians. This is a delicate matter, but it cannot be ignored. Even if you have some inner, private views that are less than flattering about politics and politicians, keep them to yourself while working with Congress. It is the kiss of death to be perceived as having a "holier than thou" attitude or, as one staff member put it, to "convey that the purity of the scientific profession puts you 'above all of this.'" Suggestions by scientists or engineers that, in the words of one Capitol Hill aide, "we are stupid and will believe anything given to us" do little to advance their cause.

5. Perform good intelligence gathering in advance. Intelligence gathering does not mean calling in the CIA, but it does involve learning at least the basics about the member, committee, or staff member you are contacting. As one staff member exclaimed, "Can you believe this person didn't even know which party my boss belongs to?" Staying in touch with developments is an important part of gathering intelligence. A good "one-stop" source is *Congressional Quarterly's Politics in America* (see the Appendix for more information); other sources include hearing records, speeches, floor statements, and conversations with Washington friends who are knowledgeable about Congress. (The Appendix lists sources of information.)

6. Always use a systematic checklist technique. Whether it be for participating in an elaborate meeting, presenting a statement in a hearing, or making a simple telephone call, prepare carefully. One good way to insure careful, complete advance preparation is to write out a good checklist. Such an approach will help you to plan better and to track your progress during a meeting or a telephone conversation. You are much less likely to get lost or to forget an important point.

7. Remember that timing is vital. All too often the message may be great, but it is useless if the timing is all wrong. Keen judgment is required here. Coming too late can mean the legislative train has left the station. As one committee staff director put it, "It was a good set of suggestions, but we'd already reported the bill out of committee two days ago. They thought we could fix it on the floor. Well, maybe — sometimes. But they should have come three months ago when it was still in subcommittee." On the other hand, coming too early can be just as bad. A good effort can be wasted "if it is too early and other matters are dominating the legislative agenda. We only handle so many things at a time," according to a senior staff person.

8. Keep the congressional calendar in mind. While activity in the congressional environment seldom comes to a complete halt, it does vary over the course of the year. A member observed, "There is a much better chance of having an in-depth discussion with me during a recess period, whether in person or on the telephone." This advice applies to staff members as well. Of course, this consideration must be balanced with the overall timing question discussed earlier. (The typical congressional calendar is outlined in the Appendix.)

9. Understand congressional limitations. It is important to have a good understanding of just what Congress can do and what it cannot do. A committee staff director said, "We don't have big planning staffs that can sit down and spend days analyzing what somebody drops in our lap — such as a 10-page memo with 45 appendices." A recurring theme is that too many people bring problems to Congress and "look to us to devise a solution instead of presenting a plan for us to consider, modify and perhaps adopt." The congressional environment, as described in Chapter 2, is one of enormous time pressures from multiple competing interests. It is not a place for much original analysis and extensive research. Bear this in mind in your contacts with Congress, whatever their type or format. Do not expect members or staff to have deep familiarity with specific pieces of legislation, or to know their provisions or even their bill numbers. You will lose them if you if you toss out statements like "Section 222 of Title III of H.R. 4494 will kill us." While it is important to be concise, it is also important to make clear what you are taking about.

10. Make it easier for those in Congress to help you by focusing your problem or issue clearly and making apparent

what decision is needed or what action Congress should take. Too often scientists and engineers who come to Congress for assistance do not make clear what the problem or issue really is. Work carefully at honing your request or advice or information so there is no doubt about your issue, your position, or what you are asking for. Do this by working out a proposed answer to your request or by presenting a plan of actions that might be required. Occasionally this might be seen as presumptuous, but more often it will be seen as helpful and well organized on your part. Members and staff appreciate proposals for action that are clear and articulate, and show that they have been thought through before presentation. Congress, if it moves on your proposal, will often use your specific bill or report language. Have the material ready to use!

11. Remember that members and staff are mostly generalists. While most are "quick studies," you cannot assume that they will immediately understand or appreciate the value of what you are proposing. Keep messages simple; do not be too detailed; and do not overwhelm your listeners with technical jargon.

12. Keep the "bottom line" in mind. In whatever way you are working with Congress, never forget for a moment what your objective is. Make it clear to them as well. If you have a hidden objective or agenda, this is not the book for you. Go back and read Machiavelli's *The Prince* instead.

13. Use time — yours and theirs — effectively. Members and staff are keenly aware of the value of time and resent having it wasted. Plan your efforts in detail and try to make your presentation as concise as possible. Coming across as disorganized or long winded (on the telephone or in writing as well as in person) is a good way to limit your future congressional contacts.

14. Don't patronize either members or staff. Even if it is clear that the person with whom you are dealing is uninformed or misguided, keep your cool and maintain a steady course. Do not resort to an "I'll show this idiot" attitude. On the other hand, it is not necessary to accept rudeness or insulting behavior meekly. While not frequent, instances of such conduct do occur. A call or a letter to a member or chairperson is one way to respond. In an extreme case, a letter to an editor is another. Finally, there is always the "Hill grapevine," which can be available through friends, association offices in Washington, and media individuals who cover Congress.

> "For effective meetings, be completely prepared about the issue and know something about the member you are seeing."
>
> REP. RAY THORNTON (D-AR)

15. **Don't underestimate the role of staff in Congress.** While it is very important to remember that members are elected and staff are not, staff members generally play powerful and influential roles in the congressional setting. Do not make the mistake of looking down on a staff member or underestimating his or her ability to help or hinder you, even if the person happens to be very junior.

16. **Remember your friends and thank them often.** These are more than simple courtesies; they are also the hallmarks of polished professionals. Keep track of your advocates and look for ways to express your appreciation. Use *handwritten* notes to stay in contact. Private thanks are sometimes appropriate, but also look for public ways to thank them for their contributions.

17. **Finally, remember that the great majority of members and staff are intelligent, hard working and dedicated to public service.** If you approach working with Congress with a positive outlook based on the recognition that on the whole they are competent and dedicated, the outcomes of the experience are much more likely to be favorable and fruitful. Members need and want your help: make it easier for them to use it more effectively.

MEETINGS

SCHEDULED, FORMAL MEETINGS WITH MEMBERS AND STAFF

One of the most frequent ways in which people from outside work with members and staff is through scheduled, formal meetings. Such meetings, which may involve either individuals or groups, are generally held to discuss specific requests, ideas, pieces of legislation, proposals, and the like. The rules laid out in the previous section will give you a good start in getting the most out of a meeting with a member or staffer. However, you should also bear in mind a number of more specific pointers.

For all meetings, but particularly for meetings with members:

1. Be on time. Being on time is a cardinal rule. It is the member's (or staffer's) turf, and most likely you asked for the meeting. Be aware that the chances are good that you will have to wait when you get there — even for a meeting with a staff member. Much goes on in the congressional environment that is beyond the control of individual members and staff. Although some members and staff are notorious for not holding to their schedules, most make a serious effort not to be late with their appointments. Build the prospect of delay into your schedule; do not take it personally or get upset. Use the time to relax or chat with a staff member who offers conversation. On the other hand, do not interrupt a clearly busy staff person or an overworked receptionist trying to cope with ringing telephones.

2. Make clear why you are there. Describing a meeting with one group of scientists, a senator said, "They were with me for twenty minutes, and when they left I still had no idea why they had come to see me." Avoid this mistake — get the problem or the issue and your request on the table right away.

3. Keep your message simple, focused, and short. Do not waste your opportunity or their time. In the words of one senator, "Time is of the essence. Make your best case quickly and up front — and let the rest happen." Even at the risk of some oversimplification, be concise. Do not waste time on too much background. And do not overwhelm them with details; instead, highlight the key facts. If they want more detail, they will ask for it.

4. Prepare carefully in advance. Put together a simple, clear summary paper on one or two pages — including a brief background — that can be left after the meeting. Know what you want to say, and know when you have said it. Keep track with a short checklist. Practice in advance with a "dry run" of your presentation. Demonstrate by what you say and how you present it that you are well organized and worth listening to.

5. Do your homework on individuals. For meetings with either members or staff do some background information gathering. For members, learn where they come from (state/district), their committee assignments and professional backgrounds, where they stand politically on various issues, and how they fit into the congressional power structure. Try to learn if the member already has a view on your issue. As one senior staff person said, "Know what is on the Member's mind in terms of recent concerns. Check recent hearings and floor debates." For staff, there is less published information but it is still possible to get reasonably accurate profiles by making a few telephone calls to Washington friends, agency staff, association staff, and the office of your own senator or representative, and by consulting the *Congressional Staff Directory* (See Appendix.).

6. Do your homework on the issue or problem. It is obvious that you should know the technical side of your issue. Not as obvious, perhaps, is the importance of translating your message into terms relevant to Congress. Know which bills (if any) are pertinent. Know which committees are involved and what they are doing (e.g., holding hearings, planning hearings, or holding the issue in "deep freeze"). Know which other members are involved and what their views are.

7. Tie your issue to member interests if possible. In doing your planning and homework, look for connections between your issue and the member's interests. Such connections might be his or her legislative interests or perhaps they might be related to constituent concerns. One senior staff person said that, although it might not always be possible, you should try to "say why your proposal is important to the Member's state or district, how current efforts are helpful that way, and why your proposal would be good for the member's constituents." More generally, seek commonalities between your interests and those of the member. You are on your way if you can clearly show the member how he or she can gain by going along with you. Finally, as one staff person said: "Try to have something to offer — good advice or useful contacts for additional information, for example."

8. Think in terms of providing a basic education. Beyond any specific objectives you may have in asking for a meeting with either a member or staff member, one senator suggests that you not forget a purpose of mutual value to you both: "Think in terms of providing a basic education about your issue and its broad context to larger congressional concerns. Do this as clearly as possible." Putting the issue in "broad context" does *not* mean a lengthy background review. It means tying your issue into one or more of the handful of large concerns facing Congress at any given time — e.g., the budget deficit, health care, the state of the economy — and doing it clearly and concisely.

> "Congress needs to learn more from scientists and engineers. You should establish ongoing relationships with members of Congress; we need to hear from you on a wide variety of issues that require scientific and technical advice, not just when you have a specific request."
>
> **REP. DON RITTER (R-PA)**

9. Encourage and be prepared for questions. Organize your presentation so as to allow for questions and discussion. One representative said, "I expect to ask questions and I like straightforward answers." A senior staff person advises, "Give a short, clear answer first — and a long answer if the circumstances lead to developing the latter. Only add details and qualifications with encouragement." One staff member has some unusual advice: "Consider the Members as rather bright, intelligent students who are not terribly well-informed on your issue."

10. Be honest. If you do not know the answer to a question, say so. A staff member says, "Don't pontificate" and don't ever "fake it" with a guess or a confusing nonanswer. If pressed, you might speculate and label your response appropriately. Another staff member advises: "Be open to all questions even if you think they are stupid or ill-informed — or reflect the views of your opposition." If you cannot do something, say so. Enough people violate this rule to cause members and staff to underscore how strongly they feel about trusting what a person says. Remember, members and staff work in an environment where one's word is one's bond. Do all of them follow this principle? Of course not. But violators of the principle — members and staff — can end up as "damaged goods," lacking the respect and trust of others. Do not misrepresent either your own or your competitor's positions; it will eventually come out. A related point suggested by a number of staff members is "Don't oversell your case." Work hard at building your credibility; it is a tremendous asset — even if your issue is weak or unpopular. To further enhance your credibility, acknowledge as accurately as you can those who disagree with you or are opposed to what you are suggesting; tell the member or staff person as best you can why this is so. Do not make them research this information or be surprised by your opponents. Also, your credibility can sometimes be enhanced by saying "I don't know" if you do not. A "know-it- all" attitude can damage your case.

11. Be adaptable and flexible. The congressional environment can be somewhat chaotic at times. In practice, this means that even if your meeting has been scheduled for weeks, it may start late, and it is subject to interruption for any number of reasons — floor votes, committee votes, or telephone calls from other members on urgent matters. Accept such interruptions gracefully. Do not be flustered by starting and stopping. Think in advance about how to pick up the threads of the conversation and weave them into your next point. Watch and listen carefully to see if you have made the transition successfully. A very quick review of your earlier points may occasionally be necessary, but do not even think of repeating everything.

12. Use concrete examples as often as possible. Congressional people tend to be oriented to examples and anecdotes rather than abstractions or broad generalities. Play to this as much as possible without distorting your case. Staff members are always looking for "nuggets" to put into member speeches, floor remarks, and committee hearing remarks. Help them out if you can. As one staff member observed, "Concrete examples seldom hurt and most often they help."

13. Be especially careful in planning group presentations. It may be ego gratifying for every member of a group to have some part in a presentation (or there may even be a valid technical reason for this), but group presentations should be used sparingly, with caution and careful planning. One person must be in charge and manage the individual presentations smoothly but firmly. A staff director advised: "Plan carefully who is going to say what. Don't have a confusing scene in the Member's or staff member's office about what is going on. Don't get into side-discussions within the group. And think about this mathematical fact: three people cannot each give a ten-minute presentation in a 20-minute appointment. This seems obvious, but it is tried often enough to boggle your mind."

It is even more important for a group to have an advance dry run than it is for a single presenter. Using a detailed checklist is also important. One long-winded presenter can ruin your entire show by forcing a carefully crafted closing of five minutes to be done in one — with the resultant loss of focus and force in achieving your objective.

14. Consider and offer appropriate follow-up. It is seldom that a single meeting with a member will be all that is necessary to achieve your objective. Possibilities range from a simple follow-up telephone conversation or two with a staff member to an extended period of working with staff. Conceivably, other members might become involved. Take this into account and be certain that follow-up commitments can be met before you offer them. In any event, before you leave any meeting with a member, try to have clearly identified the name and phone number of the staff person who will be your principal follow-up point of contact. Finally, it is useful and appropriate to ask such a staff member if he or she thinks you should contact other staff members about the issue.

In addition, for meetings with staff members, consider the following special pointers:

15. Remember that staff members can be powerful and influential. They often serve as "gatekeepers" and can help you communicate with members, if this is necessary. Do not underestimate their value by thinking you have been passed off to an underling if a member did not agree to your request for a meeting. Alienate a key staff member at your own peril; he or she may hold the keys to future contacts. In the words of a committee staff director, "Staff members have a lot of discretion on who sees Members, on who testifies at hearings, on what is read by Members, and on what goes into legislation in the form of specific words and sections." In setting up meetings with members, it is often useful to enlist the aid of a staff member before approaching the member.

> "Since everyone in Congress is pressed for time, be direct and concise. Do not go into a meeting without having a clear idea of your purpose and the main message you want to convey."
>
> **REP. RICK BOUCHER (D-VA)**

16. Remember, though, that not all staff members are equal. This is a basic fact of life in Congress. Acknowledge it and use it effectively. As part of your intelligence gathering, try to assess the roles of those staff members who appear to be important to you. Include in this assessment member–staff relationships. Some staff members have wide latitude in speaking for members; others are not authorized to tell you the time of day. Some staff members can directly influence multimillion- or multibillion-dollar programs; some simply gather information for others to use. A senior staff person advised: "Make sure you are talking with a person who can really do you some good." However, a companion to this advice is not to be upset or put out if you end up with a less than influential person; there is always another day and people do get promoted! Finally, as another senior staff person suggests: "Don't confuse staff interest in a meeting on your project or issue with automatically leading to a favorable outcome for you." Remember that, as one staff person pointed out, "most difficult decisions are made in Member offices or in committee mark-ups by Members."

17. Distinguish between personal staff and committee staff members. While a personal staff member can be the conduit for you to see a member, and he or she can serve as a liaison to appropriate committee staff members, a personal staffer will less often be the focus of major legislative or budgetary activity, especially in the House. Committee staff are likely to be more legislation-oriented or technically informed than are personal staff. Personal staff are likely to be more member oriented and district or state oriented. All of this means you must do different kinds of homework and plan your meetings accordingly.

18. Remember that time is valuable for all staff members. A senior staff person cautioned, "You need to remember that staff is generally overworked, is nearly always pressed for time, and generally handles many issues besides the one you are interested in. While interested, they may not have the level of zeal for your project that you have." In general you can expect to have somewhat more time with staff members than members. But this does not mean you can overload them with details or stacks of paper. It is often useful to have visual ways to make points quickly and effectively. One staff member said, "I look for good ways to brief my boss quickly." Another noted, "you can probably expect to meet with a person somewhat more conversant with an issue than is the Member. You may even have the opportunity to discuss philosophical underpinnings more extensively." But a bedrock theme from all staff is that severe pressure on time colors everything they do, including meetings.

19. Include even small details in your planning checklist. In preparing a checklist for your meeting, think carefully about what you will want to leave at the meeting. This should include, according to a senior staff person, your business card with the correct information — including today's date and a brief mention of your meeting topic. It was pointed out that "hundreds of cards get collected, at meetings, at receptions, and so on, and it is easy to forget how and why one has a particular calling card six months from now."

INFORMAL VISITS

Not all circumstances lend themselves to formal, scheduled meetings. Sometimes it is useful to visit with a member or a staff member for purposes more limited and less formal than a presentation or a request for action or help. While many of the points listed above apply to such visits, there are also some additional items worth highlighting.

An important purpose in making visits is the part they can play in building and maintaining good working relationships with those in Congress. Stated a little differently, visits can be part of avoiding the mistake of not being seen until you have a problem and need help. Please do not misinterpret this advice. It does not mean that first-time visitors or requests will not receive a fair hearing. Negative comments by members and staff about those who are "invisible until they need help" tend to be aimed at organizations and individuals who make a practice of this.

> "Too often, presentations are unfocused; there is not enough information on what is the problem, what is the proposed solution, and what decision is required."
>
> REP. GEORGE E. BROWN, JR. (D-CA)

Overall, the matter of visits calls for finesse and good judgment. On the one hand, relationship building is important, but on the other, you certainly do not want to be seen as a "pain in the neck" who hangs around an office wasting time or who stops by to "kill time" while waiting for an appointment elsewhere. This kind of behavior can alienate staff members and can virtually ensure that if you ever need help, it will be difficult to get it. Consider the following in planning informal visits:

1. **Call in advance to schedule most visits.** Calling in advance for most visits will keep you from wasting your time and is preferred — if not generally required — by both members and staff. This is simply good etiquette and will avoid creating unnecessary irritation.

2. **While you can consider brief "stop-by" or "pop-in" visits, do so with discretion.** As a corollary to the preceding point, it is not always necessary that every meeting or visit with a member or staff member be scheduled days or weeks in advance. Occasionally — and on a selective basis —brief "stop-by" or "pop-in" visits may be used. Either with an advance

telephone call or on an impromptu basis, your visit should be advertised as a short, one-minute-or-so affair. Such visits can be for a simple "hello — I'm in town and hadn't seen you in some time," or passing along some favorable or valuable news, or to provide an advance copy of an important report, or for saying that a report has come out or is coming out and asking, "Would you like a copy?" However, "stop-bys" or "pop-ins" are seldom, if ever, to be used unless you have some relationship with a staff member who may place value on the connection you represent. So be careful about getting yourself into an awkward or embarrassing situation. For a member, unless you are a visiting constituent or have a good personal connection (e.g., a role in a campaign), forget "pop-ins," at least in Washington.

3. Don't forget that senators and representatives have district and state offices. Often this is your best opportunity to meet with a member and command a decent attention span. It was frequently observed by members and staff that "too many people don't understand that Members can be seen in their district and state offices." As noted by one senior staff member, "This is where you can walk in and see my boss without an appointment. A lot of Members hold 'open-office' sessions for their constituents." More and more members have divided their staffs between Washington and their districts or states; this should be considered in asking for appointments. Also, personal staff members from Washington visit their members' district and state offices from time to time and may well be available there. Not to be forgotten, however, is the overwhelming orientation of district and state offices: constituent service. (See the last section of this chapter for more advice on working with district and state offices.)

4. Visit as a constituent. As just noted, constituent service is an important part of congressional life. And, as a U.S. citizen (unless you live in Washington, D.C., or on one of several islands), you are a constituent to one representative and two senators. In addition to a visit to a district or state office, you might consider visits to the Washington offices of your members if you are in town as a witness for a hearing or on other business or even on a holiday visit to the nation's capital. Even in this case, though, a phone call ahead of time is simply good manners. You may or may not get to see your members, depending on their schedules, but you can almost certainly count on seeing a personal staff member. If you are in for a hearing, take along a copy of your statement to drop off — preferably with a summary. If you do this in advance of your hearing appearance, it is not unheard of for a member to come to the hearing and introduce you. Do not count on this, however; it will depend on a number of things, including the member's interest in your topic, your status, and the member's schedule.

5. Plan all visits — even "drop-bys" — carefully. All of the advice given on planning carefully and preparing for meetings and presentations applies equally to making visits, even short ones. And as a specific reminder, even for a short visit, prepare a written checklist for yourself, including what to do before you get there and the points to be covered in your visit.

MEETING ADVICE WRAP-UP.

On any given day, members and their staffs will be talking with many different people on many different subjects in all kinds of meetings. Under such circumstances, it is easy to be forgettable or, even worse, recalled in terms such as "don't let that jerk in to see me again." You are in a severe competition for time and attention. However, people who come across as knowledgeable and credible, who know their message well and can present it with clarity, who make their request or offer simply and concisely, and who generally make it easy for the member or staff member to help them are such a rarity that they will be remembered, helped if at all possible, and called upon in the future.

USING THE TELEPHONE EFFECTIVELY

Use of the telephone dominates the congressional environment. In the fast-paced, verbal setting of Congress, this is one of the preferred modes of communication for both members and staff. It is important not only to be aware of this preference, but also to be sure you can use the telephone skillfully. Booklets and brochures on good telephone techniques and manners are a useful start. Beyond these, however, there are number of important points to keep in mind.

1. Remember, there is generally an openness in Congress to telephone contact. In the survey conducted for this handbook, dozens of staff members mentioned their general openness and receptivity to telephone contact. In their own words, staff described themselves as "always open," "frequently open," "very open — unfortunately for me," "yes, I'm here to serve the public," "as time permits," "generally open — once the relationship is established," and "will usually talk with about anyone, but realistically I'm more open to calls from people or groups known to me." One staff person said, "I return all of my calls eventually but I acknowledge that not all staff members do so — particularly if they don't know the caller."

> "It's important that you make clear what your priorities are. Nothing weakens a presentation more than giving the sense that a particular item is the most important thing you want — except for all the other items that are also the most important things you want."
>
> — **REP. ROBERT S. WALKER (R-PA)**

2. Bear in mind that members are not generally as open to telephone contact as staff. The readier availability of staff should be no surprise. However, many people do call for a member and are upset when it is not possible to be put through or when they are referred to a staff member. Representatives and senators as well as staff vary considerably in their openness. Responses to our survey revealed that many are quite open to constituent calls but that for others, as one representative noted, "it depends, it varies a lot." Discovering who is available and who is not may well take some trial-and-error searches and should be regarded as part of your intelligence gathering.

3. Don't be put off if you are referred to a staff member. Understanding the reality of working with Congress involves a recognition of the special role played by staff. This applies to telephone calls as well as meetings. Also, within the staff hierarchies it is more difficult to reach some staff members than others. As one staff director described his situation: "I try to answer many of my calls myself, but often I have to turn them over to someone else to answer for me."

4. Plan your telephone call in advance. By this time you may be tired of seeing "plan in advance," but it is hard to overstate the point. It truly upsets congressional people to get into a telephone conversation with someone who, as one staff person described it, "wanders all over the barnyard looking for the chicken." Just as you should for a meeting, prepare a little checklist even for a telephone call. It will help you to avoid the problems of wandering around trying to find a way to end the conversation or of conveying confusion because you're not sure you covered everything you intended.

5. Identify yourself immediately. The first order of business is to get your name and organization out on the table. If you have a personal connection with the individual for whom you are asking, use it. Very often, you will be going through a telephone receptionist who is very busy, so have this preliminary information ready in concise form. If you are put through, or the person answers directly, make sure, as one staff person suggested, "that it is not a bad time and say that you would like perhaps five minutes or so."

6. State your business immediately, quickly, and clearly. As in a meeting, stating your business clearly and quickly is essential. One staff member put it this way: "Remove the wonder of why this person is calling me. Don't make me guess at your motivation or why you are calling." Another staff person advised, "Don't obscure your message because you're embarrassed about asking for something. Come right out with it." Still another said, "Unless you are a personal friend, don't waste time with chitchat." On a comparable theme, another recommended: "Be concise; be specific, even if it is just to say hello." Others said: "Please, no long backgrounds. Get to the point. Know what you want." "In addition to many visitors, I have 50 to 100 telephone conversations a day. I don't have any time for long-winded baloney." "Don't get yourself to the point where the staff person is thinking, 'when will this turkey get off the phone?'"

7. **Give careful attention to whom to call and what to ask for.** Insofar as possible, do your homework before making a call, although it is quite acceptable to call someone for advice on just who is the right person to talk with. Do not waste the time of a busy staff person asking for routine information on legislation or other matters that is readily available in databases or published sources. In our interviews, staff members complained about the number of calls they get that fall into this routine information category.

8. **Avoid using telephone calls for complicated subjects.** It is important to remember that not every subject or problem is appropriate for handling on the telephone. Typical staff member advice was "if the issue is complicated, make an appointment. Don't even try to use the telephone." Still others, despite the general aversion to written forms, said they preferred a letter in advance of a telephone conversation. An admonition was to allow a reasonable time for the letter to be read. Still others suggested they preferred a letter in advance if they did not know the person. A senator said: "I respond best when telephone calls are preceded by a written communication introducing the person and the subject."

9. **Consider the referral approach.** The importance of tying in quickly any personal connection to the person you are calling was mentioned earlier. Members and staff both noted that referrals from mutual friends or close associates can be helpful in smoothing the way for a telephone call or a meeting.

10. **Be patient, but persistent.** Remember that members and staff are under horrendous time pressures. As one staff person described it, "We may not be able to return calls immediately or even the same day. And some staff return calls late in the day, so you may want to consider leaving a home number." Others observed there was nothing wrong with being persistent if you have not heard back after several days. Just try again, with politeness and aplomb.

11. **Remember that telephone calls based on "outrage" and "demands" don't go over well.** As a citizen you can exercise

your First Amendment rights to call anyone in government and tell them how rottenly they are running things. However, such calls seldom lead to anything unless they are part of a massive grass-roots outpouring on an issue that touches a raw national nerve. And if your call is made in the form of a demand, remember, as several staff members pointed out: "Rudeness doesn't help. And I don't respond very well to *demands* that I do something."

12. Consider faxing as an alternative. The "age of the fax" has arrived, and sometimes a brief fax message is far more useful than a phone call. The person you are trying to reach can respond to a fax without playing "telephone tag." More and more congressional offices are becoming fax oriented.

13. Follow-up notes can be helpful. As noted in point 8, above, some in Congress prefer a written notification in advance of a telephone call. It is often difficult to judge whether you should write in advance or just call. However, depending on the nature of your telephone conversation and how it went, you may wish to consider a follow-up note in a letter or a fax. It may be as simple as a "thank you" or a bit more in the form of a summary of the points made during your conversation.

TELEPHONE ADVICE WRAP-UP

Since the telephone is a dominant feature of the congressional environment, learn to use this communication system well. While Congress is open to telephone calls, do not abuse this openness with trivial and poorly planned calls. Prepare in advance with a checklist. Identify yourself — with a referral if possible — and promptly get to the crux of your business. Never be rude or demanding; politeness, patience, and persistence will pay off.

PREPARING AND SUBMITTING CORRESPONDENCE

Although use of the telephone is one of the defining characteristics of Congress, letters and reports are also a necessary feature of congressional business. In fact, it is not uncommon for a member or staff member to request a letter under a variety of circumstances: to ask for a meeting; to ask for a telephone conversation; to describe a problem that calls for some action; and to provide information. The emergence of fax technology has probably shifted some focus to putting things in writing. But in this guide, no major distinction will be drawn between correspondence by mail and correspondence by fax. Only a general admonition will be offered: do not let the ease of faxing beguile you into careless planning and preparation. Maintain the same high standards in a fax as you would use for a traditional letter.

Given that correspondence remains important, if not preferred, it is important for you to understand how to use this mode of communication effectively. Here, then, are a number of ideas that can help.

1. Remember that letters can be useful. Although they are burdensome to handle at times, especially when they pile up day after day, staff members admit that some letters are helpful. "To be effective, however, they must be written with care," said one staffer. "I hate trying to wade through prose that sticks to your brain cells like mud on your shoes." And it is important to recognize the limitations of letters, according to other staff, in that telephone or personal follow-up is often required.

2. Remember that letters need not ask for something. Among the most useful letters and reports are those that provide information in easy, readable, and understandable form. For example, as one staff member observed, "I find them useful if they contain constructive comments on pending legislation, especially if they represent consensus positions taken by reputable groups." Another added, "they are valuable to me when they point toward issues of mismanagement, etc., that are not widely known already." Oversight and investigative hearings have actually resulted from letters of this type.

3. Make your letters short and to the point. Brevity is a word that comes through often in talking with members and staff about correspondence —both letters and reports. As one staff member said: "Volume kills! An A- Number One letter for me is one page." Another advised, "Be specific. Don't make me guess what the point of your letter is." A clear, concise, well-reasoned presentation of an issue, problem, or request is what is desired. Getting to the bottom line quickly was the advice of one senator: "Tell me what you want or have to say in a nutshell." On this point of focusing your letter, a senior committee chairman said, "Put the details in an enclosure." Other advice was not to clutter your message with an overabundance of facts. Keep them short, selective, and pertinent. A senior staff person advised: "Don't be coy or beat around the bush. Get to the point. Be up front about what you want."

> "We take a lot of time to understand science; please take some time to understand us. Keep in mind you are dealing with focused generalists, not narrow specialists; get to the point in an understandable manner."
>
> **REP. SHERWOOD BOEHLERT (R-NY)**

4. Demonstrate familiarity with congressional processes. Demonstrate that you have done your homework and know how your issue fits in. If there are various views on an issue, make clear that you are aware of these. It is even better, according to staff members, if you can provide a brief summary of your opponents' arguments. As long as you are fair and accurate with your facts, there is nothing wrong, as one staff person said, with "telling us why the others guys are wrong."

5. Don't assume familiarity with your issue. This admonition of not assuming familiarity may seem at odds with the advice on brevity and conciseness. It is not. It simply means you must strip down your case and your data to what is absolutely necessary to communicate your main objective. Your letter should be a model of simple elegance, in which the reader can flow easily through your prose despite not being the expert you are.

6. Make the case for congressional action or interest. It is not always clear to a member or staff member just why they should take time to become involved with you or your issue. Part of your letter must be devoted to making this case concisely. It is also important for you, as one senior staff person advised, "to describe what else has been done — that is, what you have done to help yourself, other than to contact Congress."

7. Make clear who you are. It may seem obvious but it is essential that you be clear as to what role you are playing in sending a letter or report. And a little personal information is necessary if you are unknown to the proposed recipient. Clearly indicate your addresses, telephone numbers, and times you are available. One staff member noted, "It's surprising how small things like this are not attended to properly. It makes it much more difficult to follow-up."

8. Be wary of using mass mailings or form letters. A good way to raise the blood pressure of a staff member and even members who are used to this sort of thing is to mention "mass mailings." They are very controversial. On the one hand, as various staff people said: "I ignore mass mailings and form letters!" "I've been a victim of mass mailings and I don't like them." "I don't pay much attention to 'tear out card' mass mailings." On the other hand, it was acknowledged by others that "mass mailings — when done with care — can be effective in getting attention." While orchestrated lobbying efforts through mass mailings are not generally effective, they clearly have worked in some circumstances, according to various staff members. The advice is to be careful in using the mass mailing approach — not to avoid it entirely.

9. Tie your letter to a member's district or state interest. As one committee staff person said, "From a crass point of view, it helps if the letter comes from the Chairman's district." In this vein, some in Congress particularly like letters conveying "political information" — such as "I'm upset about . . ." or "This program is a problem because . . . ."

> "While you should rely on your expertise in providing advice to us, don't become totally focused on your specific point of view. The job of a senator is to understand and balance a variety of interests and priorities in searching to serve the general public interest. So help us by keeping your eye on how your concerns are related to this general interest."
>
> **SENATOR BILL BRADLEY (D-NJ)**

10. Don't seek to become pen pals with staff or members. Some staff and members will shudder and show voluminous files from "pen pals" who spend much of their time writing letters to the office, or so it seems. Do not take a friendly response to your letter as an invitation to undertake an ongoing correspondence. It will require much more than this to develop a real relationship. In fact, sometimes it is most effective to craft your letter in such a way as not to require a written response. One staff member put it this way: "I know it may sound silly, but it eases dealing with a letter administratively if a telephone response can be made."

11. Think about your addressee. If you are writing to an office for the first time, it is probably best to address the member. Subsequently, as an issue or relationship develops, it may be appropriate to address the designated staff member. In any case, most members do not see much mail.

CORRESPONDENCE ADVICE WRAP-UP

Remember your goal in preparing the correspondence: to get it read and acted upon. Make it easy to read. Organize your material with care. Work to make it concise and clear. And test it on others before sending it. Individualized letters probably have the best chance of being read — as compared to a form letter or tear-off card. Letters that are polite and politically realistic are better received than polemics. You may even get an accolade that you might not ever hear about: "Hey, would you like to read a really good letter," one staff person calls to another. More to the point, your letter could well join the select stack of letters that go to a member for a personal look and a response to a staff recommendation.

WORKING WITH STATE AND DISTRICT OFFICES

State and district offices offer a relatively easy way of gaining access to members who may be difficult to see in Washington. In our survey, however, staff members repeatedly mentioned how few scientists and engineers seemed to be aware of this channel. There are several things to keep in mind.

1. Take advantage of walk-in appointment periods. Many representatives and senators have "walk-in" appointment periods in their state and district offices — although this does not mean you should just walk in without any preparation. Find out where these offices are located and inquire about the schedule for such periods. You can obtain locations and telephone numbers through your local library or telephone directory, or by calling the Washington offices of your representative and senators. Send a note or a fax ahead of time stating your purpose. Since these practices vary among members, you must determine how your own members operate: when and how long the time slots are, whether you can use any presentation devices, and so on.

2. Issue-oriented staff are more often located in Washington offices. The staffing mode for most congressional offices involves placing legislative and budget staff members in Washington. The state and district offices are more likely to be staffed to handle a variety of constituent services. Yet this does not mean that you cannot discuss your issue with a local staff member. For example, you can get advice on just whom to contact in Washington or on the feasibility of meeting with the member at a future date. However, unless your issue is "hot" locally, you could well be referred to the issue person in Washington. Even so, the district and state staff will usually make every effort to be helpful.

3. The guidelines and pointers for meetings and visits in Washington apply in district or state offices. Meeting with the member in the district may be more relaxed, but that does not mean you should be any less prepared than you would be for a meeting in Washington. Do not assume that being a constituent permits you to hold to a lesser standard of excellence in your remarks or presentation. Focus on making it a memorable and valuable experience for the member as well as for yourself. Even here, you must "rise above the clutter." This includes taking advantage of the opportunity to relate your concerns to local issues and to give examples of how your colleagues might feel about the issue you are addressing. Tying in other constituent interest lends support to your position. (You have a special burden to represent such views as accurately and fairly as possible in asserting to the member that you are not the only constituent who has an interest in the issue.)

4. Arrange meetings for your member with local chapters of scientific and engineering societies. While working in their districts and states, members usually have a full schedule of meetings and appointments. You can work with the local offices in requesting the participation of your members in local chapter or regional meetings of your societies. If you give enough notice and have some flexibility in your request, members say that the chances are good that they can appear. As a practical consideration, do your best to assure a good turnout. Spending an evening talking to five or six people may not be the most effective use of a member's time. When you "sell" your meeting, give a reasonably accurate forecast of how many will be there.

For example, one senior staff member reported: "My boss had to cancel out of a district event at the last minute because of a late floor vote and I went in his place. They had told us there would be a hundred people, but due to faulty publicity, it ended up with less than a dozen. Fortunately, I did some other district business so it was not a total waste, but I sure wasn't very happy with that group."

> "In general, we trust information from scientists. But to keep that trust you must clearly state which of your points is opinion, theory, or widely accepted fact."
>
> **SENATOR PETE V. DOMENICI (R-NM)**

5. Arrange visits to local research laboratories, high-tech manufacturing plants, and similar establishments. Members say they are interested in visiting district organizations where interesting work is being done or where they have an opportunity to talk with workers. As one representative said: "I like to know when a company in my district has come up with some new product or development that is significant."

6. Arrange visits to colleges and universities. One representative noted, "Over the years, we have developed good relationships with a number of researchers at universities in my district." He identified these relationships as among the most favorable of his experiences in working with the scientific and engineering communities. You, too, can work at developing such relationships: your work gets exposure to the member, and the member gets information as well as exposure to your colleagues. Building such a relationship can eventually result in your serving as an informal advisor to the Member, providing information, opinions, and perhaps more formal studies on matters of importance to the member.

7. Consider volunteer work in a member's political campaign. Apart from your working with a member of Congress and his or her staff as a constituent seeking assistance or giving advice and information, there is the special role of serving in a political campaign. Scientific and technical issues are political in the same sense that other areas of our modern society are political. Decisions entail setting priorities and allocating budgets, and these are quintessential political decisions. Yet more than a few members of Congress have had the experience of being a strong supporter of science and technology but not having this translated into political support in the same way that being a proponent of labor, business, or agriculture generates support.

This is a difficult issue to raise, but it deserves to be placed on the table. First, by no means does seeking assistance from a member of Congress involve a *quid pro quo* in the form of an obligation for political support on your part. However, there is — accurate or not — a perception in the world of politics that a sizeable number of scientists and engineers believe they are "above it all" and feel that being involved with politicians is inconsistent with the ethos of the scientific and engineering professions. While many interest groups provide support — financial and otherwise — for increasingly costly political campaigns, various members of Congress (including some defeated ones who were strong supporters of science) claim to have observed a general aloofness on the part of the scientific and engineering community.

Obviously, support for science and technology should not be the sole basis on which you make your political choice in an election campaign. But if there is a general compatibility on issues and you believe the individual has been doing a creditable job in office, then you should think about taking part in the member's campaign. This can include arranging voter meetings, organizing fundraising events, publicizing member achievements and contributions, and making financial contributions. This is not the place to describe in detail the operations and functions necessary to an election campaign, and your decision will depend on your own personal style, abilities, and interests, but the point should be clear: politics is important to science and engineering. Consider getting involved.

CHAPTER 5

HEARINGS AND TESTIMONY

The previous chapter provided pointers and guidelines on meetings, visits, telephone calls, and correspondence with members of Congress and their staffs. In one sense, these activities can be considered as preliminaries to performing in the center ring of the circus: appearing as a witness in a hearing before a congressional committee. While the reality may be that a private meeting with a powerful member or staff member can be more influential than testifying in a public hearing, there is a strong belief — inside and outside of Congress — that holding public hearings is one of the more important things that Congress does. In recent years, Congress has been averaging several thousand hearings per year, calling upon tens of thousands of witnesses.

In short, public hearings are important, and Congress spends a lot of time and energy on and in them. Yet, with thousands of witnesses appearing annually in the aggregate and perhaps a dozen or more at any one hearing, there is a distinct and difficult challenge for any witness. As described by one senior staff member, this challenge is "to rise above the clutter." By way of explanation, this person said: "We are deluged with paper and telephone calls, hearings, meetings, and visits. You have a lot of competition and you have to figure out a way to distinguish yourself and what you have to say from all the thousands of others who are likewise competing for our attention."

The following pointers and guidelines are intended to help you do just that. There is relatively little published material explicitly devoted to testifying before Congress, but probably the best single publication is *Testifying With Impact* by Arch Lustberg. Lustberg's booklet covers virtually all aspects of testifying — from breathing exercises to relax yourself to avoiding "deadly" facial expressions and other mannerisms to general advice on communicating more effectively. For those who are planning to testify before Congress, it should serve as important companion piece to this guide. (See "Suggested Readings" in the Appendix.)

GUIDELINES FOR PREPARING TESTIMONY

1. First, understand the purpose of the hearings. Regrettably, not all witnesses come to a hearing with an adequate understanding of just why the committee is holding it. In practice, this also means that you should answer — or attempt to answer — the questions asked in your invitation letter or telephone call. Staff members observed, "You might be surprised at how many witnesses don't do this." One firm admonition was to "be responsive to the subject or purpose of the hearing." In summary, you can take the opportunity to make your own pitch, but first you should respond to the needs of the committee and its invitation. Earlier advice on intelligence gathering applies here as well. Find out "what really is going on."

2. Consult with committee staff in advance. You can take steps to avoid some of the problems noted above by consulting with and working with committee staff members as far in advance of the hearing as possible. This does not mean that you must adhere to committee guidance on what views or positions you should express. But you can learn much about how best to focus your remarks on issues raised by the committee, get a better understanding of the turf and the personalities, and pick up general advice on how to present yourself most effectively. Unless you are appearing under a subpoena (a special circumstance not covered in this guide), consider the committee staff as allies who can ease the way for you. Sometimes they will tell you in advance what questions their bosses are likely to ask — because they often draft these questions.

3. Write with simplicity, brevity, and clarity. These are admonitions that run through all aspects of working with Congress, but they must be emphasized in preparing testimony. A staff member advises: "Your audience doesn't have a scientific or technical background, so write for the layperson." Another said, "Make it understandable for the Congressperson, with concrete examples if possible." As a specific suggestion, a member said: "I like to see a bullet approach with tightly written phrases. This helps me to understand more quickly what you're getting at." A senator said he found it helpful for a statement "to use imagery that can capture my attention." Still others observed that, on the whole, members are uncomfortable with abstractions. A senator said: "Write concisely; give me clear, articulate analysis; put the details in attachments." In summary, "write clearly" comes through as an overwhelming plea from members and staff alike.

> "I always find it useful if a group seeking help from Congress outlines what they are doing first to help themselves. Too often we see an entitlement mentality at work among those who lobby us for funding."
>
> **SENATOR JEFF BINGAMAN (D-NM)**

4. Place your summary up front and highlight your key points. Your main message should come through early. Put the details or backup data in appendices. While it is acceptable to provide a background summary, don't overdo it. And make sure your "vision" comes through clearly, as advised by one senior staff person. From your testimony, the committee must be able to understand, as another staff member said, "where things stand; what needs improving or changing." A senator called for a statement to highlight the most compelling data and to have a distinct "bottom-line orientation." A staff member said, "We like statements that convey facts, contain original analysis, and clearly state a position." Finally, a committee chairman said: "I want to understand what course of action you are proposing and your justification for it."

5. Make sure your facts are correct. In presenting your case, it is quite acceptable to be candid and forthright — as long as you are sure of your facts. A senior staff member put it this way: "You can be frank — but factual." It can be worse than just embarrassing if you are found to have been sloppy with your facts; if your situation calls for the use of information that needs to be qualified, do so. Where appropriate, characterize the nature or range of uncertainty.

There is a balance to be achieved here: congressional testimony is not a scholarly document, reminded a senior staff member, but it should be well documented. Avoid voluminous footnotes; brief citations are in order. Also, statistics, graphs, and charts can be desirable if they make clear points and are used with discretion.

6. Give careful attention to organizing your material. Think carefully about how best to present your testimony. Play with the sequence of your facts, suggestions, observations, and conclusions in order to find the most effective way to make your case. Seldom, if ever, will a first draft be acceptable. You must work at this process. One good way is to begin with a "stripped-down" outline of no more than one page. Organize and reorganize this outline until it comes together. A representative admonished, "Good organization is essential." It was his view that this was the way to achieve completeness while avoiding the redundancy and rambling of poor statements.

7. Comply with committee administrative requirements. There are a number of items that collectively can make or break your testimony, or at least strongly influence the way it is received. Submit your written statement in the required number of copies by the deadline requested by the committee; this will vary, but it is often 48 hours in advance of a hearing. Unless you were literally asked at the last minute, bringing your statement with you on the day of the hearing is not acceptable. Some witnesses can get away with this, but most people would be starting off with two strikes against them. In addition to being courteous, submitting testimony in advance of the hearing enables staff (and sometimes members) to review your statement and prepare questions for the hearing. This greatly enhances your opportunities for dialogue with committee members.

Check in advance on the desired format of the statement. Some committees require, for example, single-spaced testimony of no more than five pages; others do not care. In general, however, double spacing with ample margins and space between sections should be used. Many members and staff like to mark or make notes on a statement; make it easy for them to do so.

PRESENTING TESTIMONY

1. It may be helpful to think of a hearing as theater. One senior staff member provided a perceptive insight into hearings by observing that although testimony is not a speech, it must still be listened to by an audience. In his experience, it was most like theater with a specialized audience. Thus, his advice was to consider presenting a statement as a "theatrical performance" that you as the witness/actor/actress ought to prepare for just as you might for an opening night. In developing the imagery, he said, "Don't make it difficult for the audience to endure. There have been times when I wanted to jump up and shout, `Shut up and get out of here.' There have been other times when I wanted to applaud." And the sad part is, he noted, "some witnesses are so bad they don't get listened to even though their message may be very important."

2. Make sure you are thoroughly familiar with your statement. This cautionary note is offered primarily to those who may be delivering testimony which has been prepared by others. In essence, by the time you present the statement, it should be yours. Go through the statement line by line to make sure there is not some fact or conclusion with which you are insufficiently familiar. Know the main points well enough to present them without reading a single word if that becomes necessary.

3. Remember that people in Congress are generally more verbally oriented than print oriented. The predominant verbal orientation of Congress strongly affects how its members and staff function. This is why members and staff place so much emphasis on oral testimony. Yet, there are enough members and staff who are sufficiently oriented to the printed word to place what amounts to double demands on you in the hearing setting. Nevertheless, most members and staff would agree with the committee staff director who noted that "oral testimony is especially important."

4. Plan on summarizing your prepared statement. Within the past decade or so, there has been a growing trend for committees to insist that witnesses summarize their longer written statements. With few exceptions, you can expect a committee or subcommittee chairperson to announce that "your full statement will be made, without objection, a part of the hearing record." You will then be given a certain amount of time — usually about 10 minutes — to summarize your statement. The time allocated to you may be even less if the hearing is running late, as often happens. If there have been interruptions for members to leave for a vote on the House or Senate floor, you should prepare yourself for a request to make your summary very short.

5. Consider your summary as a stage setter. In a variety of ways, members and staff advise that your summary statement should be crafted and presented so as to set the stage for a good question and answer session with the committee members. As one senior staff person put it, "The idea is to get a dialogue with the members while you are there. They can read your statement while you are elsewhere if you grab their interest."

> "Members of Congress must address many, often unrelated, issues each day. Provide concise, honest background information without exaggeration to help the member focus in on your point. Be clear on how your request will affect the member's constituents positively and negatively."
>
> **REP. TIM VALENTINE (D-NC)**

6. Choose a preferred mode for summarizing your statement. You will seldom have any choice about whether or not to summarize your statement. In general, however, you will have some choice about which of five basic ways to use. Work with committee staff in making your decision on presentation style and which of the following models to select. *NOTE: It usually takes 2–2$\frac{1}{2}$ minutes to read one double-spaced page. This means a maximum of 4–5 pages for your summary.*

(a) Mark your prepared statement to highlight the way you will summarize it. Include "road-map" instructions to let the committee know where you are in your prepared remarks as you go through your summary.

(b) Prepare a separate, shorter version of your statement to be given orally in summarized form. Give copies of this short version to the committee staff when you arrive at the hearing for distribution just before you present your testimony. This approach serves the needs of those members who are more print oriented or who may have to miss part of your oral testimony.

(c) Use a version of your statement in presenting your testimony, but do not give copies to the committee. Make clear, however, that you are not using your prepared written statement. This will tend to minimize the tendency of members to leaf through their copies of your statement attempting to see where you are.

(d) Use a written one-page outline of your key talking points as a road-map to guide your presentation, making it seem extemporaneous. But it will not really be extemporaneous; you will have practiced this routine a number of times so as to present eloquently within your time limit. This approach and the previous ones are designed to keep you from wandering off track —as do so many witnesses.

(e) Unless you are an especially gifted presenter, **do not** "just wing it."

7. **Keep your oral summary simple, concise, and brief.** These are the familiar themes that permeate all dealings with Congress but they are especially important in the hearing context. You are out there in public and on the record for all the world to see. As one staff person said, "Make your points briefly and then say 'I'm open to questions.'"

8. **Inquire about other witnesses.** It is often helpful to try to find out in advance who else will be testifying and what their

key points well be. You will then be better prepared if you are asked during the hearing to respond to positions taken by these individuals. Do not be overly critical of the testimony of others. Hearings are often deliberately set up to hear opposing points of views.

9. Arrive well in advance of your scheduled appearance. If you are the lead-off witness for a hearing, plan your arrival for at least 30 minutes to an hour in advance of the scheduled start. To arrive at the last minute and begin in a flustered state can get you off on the wrong foot. Also, by arriving early, you may be able to have a few minutes with a committee staff member with whom you may only have talked with on the telephone. It is even possible that you may obtain some last-minute intelligence on the hearing that could be of value in your presentation.

Even if you are not the first witness, it is still important to arrive early and listen to the witnesses preceding you. You can get a sense of how the hearing is unfolding, what kinds of questions are being asked, and which members are active in the process. It is impressive to the committee when, later, you can refer knowledgeably in your own presentation to some earlier point made by another witness or, better yet, a member.

10. Carefully study Lustberg's *Testifying With Impact*. He describes the skills and techniques of an effective presenter and speaker. Reinforcing Lustberg's guidance, staff and members responding to our survey made such suggestions as the following:

Speak clearly.

Avoid looking down at the table; seek eye contact with members. A senator said: "Talk to me."

Be direct and assertive without being overbearing; convey an air of confidence about yourself and what you are saying.

Be animated. It is the "kiss of death" to sit and read a statement in a monotone. Speak "from the heart," said a senior staff person, and do not worry about your grammar. Try to convey your message with excitement, enthusiasm, and liveliness. A senior staff person said: "Members like a knowledgeable, enthusiastic witness."

Choose a style most suitable for the circumstances. Some hearings take place more informally than do others; some chairpersons are relaxed and easy going, while others are sticklers for procedure and protocol. Plan your appearance accordingly.

> "Be succinct, be brief, and be confident. As scientists and engineers you have the technical expertise and knowledge critical to the policy debates here in Congress."
>
> SENATOR TOM HARKIN (D-IA)

Tell a story, use examples and imagery, and strike a balance: do not be too technical but do not talk down to your audience either. Avoid jargon. Be relevant. Make analogies to other important issues that — based on your intelligence gathering — might be important or familiar to the members and the chairperson.

Talk to the committee — not the audience. Liberate yourself from the printed page.

11. Be very cautious about the use of gimmicks. While the name of the game is to "rise above the clutter," it can be dangerous and counterproductive to use unusual or shocking attention getters — particularly without discussing them with staff in advance. If you are going to pass around some item for inspection, be sure it is "on point" and does not distract the committee from your main message. Novelty acts should be avoided; caution and finesse are in order. This does not mean that controversy is inappropriate or that you cannot be provocative under the right circumstances.

12. Avoid planned jokes. Jokes fall into the "gimmick" category and should be used only with extreme care and finesse. Quips and humorous twists on some event or comment during the hearing are fine if they fit and if they are spontaneous. But unless you are good at this, it is better to stick to your testimony and avoid the comedian role.

13. You can use charts and figures — but with care. In your prepared statement you can use charts and figures as you think necessary, as suggested earlier. However, in your oral summary you must be much more selective in their use, if you use any at all. Remember, said a staff member, "your time is short, so think carefully about what you might gain by the use of such techniques." Think, too, about the logistics of such items, including who will show them and room lighting. Charts and graphs must be well designed and tied into your main points, according to a senior staff member. Another observed that "busy charts are guaranteed to lose your audience." In sum, such devices can serve a useful purpose and be very effective in conveying a point, but make sure they are well designed before you decide to use them. Ask in advance if it is appropriate to bring board charts and figures, slides, or transparencies with you.

14. Be careful about going off on tangents. Think of your presentation this way: You have 10 minutes to make five points and if you spend five minutes on your first one because you have gone off on a tangent, you are in trouble. According to members and staff, this is the error most frequently committed by witnesses — especially those who attempt to summarize without good planning. But even experienced witnesses can fall into this trap as an interesting idea occurs to them and they pursue it. All too often witnesses get into the embarrassing situation of having used up their time to cover no more than half of what they had wanted to say. Strict discipline, careful preparation of whichever presentation mode you have chosen, and practice are essential to avoiding this serious problem.

> "'Compromise' is a dirty word to a scientist. To an engineer, it means trading off conflicting requirements. In politics, it is the only way to get anything done."
>
> **SENATOR JOHN GLENN (D-OH)**

15. In answering questions, remember the following:

- One reason for your appearance as a witness is to present your ideas and information to the committee. Answering their questions is a good way to do this. Work on this part of the hearing as much as on the other parts (e.g., preparing your testimony). Anticipate questions and prepare for them. In particular, work with committee staff and personal staff to identify potential areas of interest on the part of members.

- As with your oral summary, do not go off on tangents; answer questions concisely and directly. If members want more, let them follow up by asking for it.

- This part of the hearing allows you, as a member says, to draw upon your written statement as well as your experience, to reinforce your points, or to make other ones.

- If you do not know the answer to a question, say so. If pressed, give your qualifiers or ranges of uncertainty with whatever response you decide to make. Offer to provide the answer for the record, in writing, if you like.

- It is acceptable to equivocate at times with a "yes, but . . ." answer, but do not overdo it. As one senator put it, after listening to several "on the one hand . . . and on the other" statements, "We need more one-armed scientists."

- In a panel setting, you may be asked a question by name or it may come to the panel at large. In the latter case, let good manners and common sense prevail. Neither a "question hog" nor a "shrinking violet" should you be. If another witness is tending to dominate the responses and you really do have something to say, assert yourself. If a question is addressed to the entire panel, it is sometimes helpful to glance at your fellow panelists and quickly gauge who is most anxious to answer first.

- You can disagree politely with another witness or even a member, but do it with good humor and grace. Do not lose your cool.

AFTER THE HEARING

1. There are still hearing-related things to do. The hearing is not really over when the chairperson gavels the committee into adjournment after the last witness has finished. There is still the hearing record to complete and possibly a report to be prepared along with associated legislation. You, as one of the witnesses, will be involved in the posthearing process.

2. Respond promptly to questions submitted for the record. It is customary in many congressional hearings for individual members and the committee staff on behalf of the committee to submit questions in writing to a witness for later answer. While it is more common for such questions to be submitted to organizations, individual witnesses also get them. Think of this as an opportunity to expand on your written statement or your personal presentation at the hearing, including the question-and-answer period. It is possible to negotiate downward the number of questions and the length and nature of the response requested, if they seem onerous. On the whole, committee staff are quite cooperative in this respect and understand the limits of individuals. The main point is for you to respond promptly to those questions you choose to answer. You may also take this opportunity to offer additional comments on questions posed to you earlier in the hearing itself. Further, you may even suggest questions to the committee staff that ought, in your judgment, to be asked of you or others.

3. Review the hearing transcript promptly. After the hearing you will be sent an excerpt of the transcript of the hearing record. This will contain both your oral statement and any question-and-answer exchanges that took place on the day of the hearing. You will be given the opportunity to correct any mistakes in the transcription process. Although the details vary by committee, in general you will not be permitted to rewrite your testimony or to fix grammatical mistakes in order to make your remarks look better. As in responding to questions for the record, it is important for you to respond quickly to the request by the committee.

4. Follow committee action in terms of any report or legislative result. The hearing is only one part of the process, and it will be important to follow up to see what, if anything, happens as a result of it. You may want to consider additional actions in contacting Congress depending upon what happens.

5. Contact the committee staff for a posthearing critique. Even if only by telephone, you should make a follow-up contact to determine how the staff thinks the hearing went. It would be especially valuable for you to hear how your presentation and appearance were received. This type of critique will not always be possible to get, but many staff are inclined to be helpful in this regard.

CHAPTER 6

A CAUTIONARY
NOTE

This guide is intended to encourage you to become more actively involved in congressional affairs and to make you more effective in working with Congress. It should help you understand how Congress operates and give you a relative advantage over those who might attempt to become involved in congressional affairs without a basic understanding of the institution, its organization, and its traditions. The guidelines for meetings, correspondence, testimony, and the like should be of substantial use in your contacts with Members and staff. Do not be fooled, however. Neither this book nor any other will make you an instant expert on Congress. The final piece of advice we have to offer, therefore, is to **recognize the limitations of your knowledge — that is, know what you don't know and when to get help.**

First of all, bear in mind that the overview contained in Chapters 2 and 3 presents a highly simplified picture of Congress. Dozens of scholarly books based on years of research, as well as numerous personal memoirs based on years of experience have been written about Congress. A few of these are referenced in the notes to Chapters 2 and 3 or are listed among the suggested readings in the Appendix. The richness and variety of the many dimensions of Congress as described by congressional scholars, along with the many revealing insights presented in the personal memoirs, are difficult to capture in a brief overview such as this. We hope that rather than satisfying your desire for information, this guide has merely whetted your appetite. If you have a serious interest in working with Congress, you need to recognize that this is not a stand-alone source of information. You should couple your use of it with additional readings or, better yet, with help from others who are knowledgeable and experienced in working on The Hill. The best guide is a human one — a colleague who has first-hand experience in working with Congress and who can give you advice and assistance in the context of your specific situation and needs. Technically trained individuals who have worked in Congress for a substantial period of time, such as the Congressional Science and Engineering Fellows sponsored by many scientific and engineering societies, are ideal sources of such advice.*

A second caveat arises from the nature of this book, in which experience and guidance from a wide range of individuals has been filtered through the lens of the author's judgment. While every effort was made not only to reflect accurately the views of others but also to be fair when generalizing from individual pieces of advice, it may well be that these objectives were not always achieved successfully. It is likely, therefore, that not everyone will agree with the advice and opinions offered in Chapters 4 and 5. It is important to remember that the "rules" in these chapters are subject to interpretation, to exceptions depending on circumstances, and (occasionally) to out-and-out disagreements. They are guidelines and suggestions to be used judiciously, not followed blindly.

A final note on political involvement — which is both a caution and an exhortation — is in order: In the concluding section of Chapter 4 it was suggested that scientists and engi-

---

* AAAS sponsors two such Fellows each year, coordinates the Fellowship programs of nearly two dozen other scientific and engineering societies, and maintains a network and database of former Fellows. For further information contact the Directorate of Science and Policy Programs, AAAS, 1333 H Street, N.W., Washington, D.C. 20005, telephone 202-326-6600.

neers who are seriously interested in working with Congress might consider volunteer work in a member's political campaign. Among those who reviewed the draft manuscript for this guide, there were some rather strong differences of opinion about whether this topic and this piece of advice was appropriate. In the author's view, these differences reflect decades of ambivalence and differences within the scientific and engineering communities about the degree and nature of their relationships with the world of politics.

The debates over these relationships are now rather more muted than they were in earlier decades —particularly during the 1930s and 1940s — when they were fierce, public, and acrimonious.

The central concern of many scientists in that earlier period was the possible loss of "independence" on the part of science if it became too closely connected with politics and too dependent on government support. While this is not the place for a thorough discussion of this complex topic, in the author's view, it seems fair to say that, notwithstanding occasional forays by political actors into science in the form of priority-setting — especially in the biomedical fields — the scientific community has retained a great deal of autonomy in the allocation of research support within fields and in the management and control of basic scientific research.

Although this is not widely appreciated in the scientific community, this independence and autonomy have been maintained as often as not through alliances among key politicians and scientific leaders. Over the years there are numerous examples of congressional figures fending off attacks on science from other politicians and sectors of the public. While there are many reasons for scientists and engineers to work with Congress — as was discussed in Chapter 1 — self-interest alone is important enough to warrant involvement. All too often we hear scientists and engineers bemoaning the lack of scientific and technical understanding in Congress. If we, as scientists and engineers, expect Congress to understand us, it is essential that we make more of an effort to understand and work with them.

Glossary*

Appropriations: Money legally set aside for a specific use. Appropriations are determined annually by Congress for the following fiscal year for every program or activity funded by the federal government.

Appropriations bill: These bills are a special form of legislation that legally set aside money for a specific purpose. General appropriation bills originate in the House.

Appropriations cycle: Once a separate congressional process, the yearly review and approval by Congress of spending levels for the following fiscal year is now integrated with the overall congressional budget cycle. Appropriations committee hearings begin in February after the President submits his budget to the Congress and often continue through May and June. After the appropriations committees complete their hearings, mark-up (decisions on funding levels) begins. By July appropriation subcommittees report their bills to the floor of their respective houses and conference committee considerations begin. The current schedule calls for all actions on appropriation bills to be completed in time for presidential signature by September 30 of each year.

Assignment process - bills: The decision by the speaker of the House or the majority leader of the Senate determining which committee in their respective houses will handle a bill after its introduction. The assignment process often decides the fate of a bill.

Assignment process - committee: Appointment of a member of Congress to various House or Senate committees. Initial committee assignments are determined by special partisan committees in each house that recommend candidates to their majority or minority leaders.

Authorization: The act of sanctioning federal support of a program or activity. In addition to specifying what a program is intended to do and who will do it, authorizing legislation often contains a section that provides for a fixed level of funding to meet program costs. While such funds may be authorized, no money may be spent unless

---

* This glossary has been adapted from Pamela Ebert-Flattau's *A Legislative Guide* (prepared for distribution by the Association for the Advancement of Psychology, Washington, DC, 1980).

it also has been appropriated from the Treasury through the separate legislative process of appropriating funds. Authorizing legislation is handled by the House and Senate through committees named for the subject matter over which they have jurisdiction, such as agriculture, armed services, foreign affairs, housing, nutrition, and science. Authorizing committees vary in their power and influence: some have tremendous power (e.g., armed services in both the House and Senate, commerce committees); others have less influence.

Authorization bill: The form of legislation used to incorporate the provisions of an authorization process for a program or activity.

Authorization cycle: The review process involving approval/disapproval or creation/abolition by Congress of programs or activities proposed for the federal government to perform. Programs and activities may be authorized to operate for periods from one year to well over five years. At the time of its expiration, the program or activity is considered by Congress for reauthorization. Authorization hearings typically begin in February and continue through the months of March and April. Congressional authorizing committees must submit anticipated legislative authorizations to the Congressional Budget Office, and to the House or Senate Budget Committee, so that anticipated levels of outlays can be worked into the proposed congressional budget. Authorizing committees of the House and Senate must report new authorizing legislation during May and June. Congress begins floor action on such legislation during July, which may continue through August.

Beltway: The circumferential highway around Washington, D.C. Used to differentiate between things inside and outside of Washington.

Bicameral legislature: Comprised of two equal houses. Congress is a bicameral legislature comprised of the House of Representatives and the Senate. Each house sets its own rules and proceedings.

Bill: One of the four principal forms used for legislation; besides the bill, there is the joint resolution, the concurrent resolution, and the simple resolution.

Briefing: A meeting in which a member of Congress receives information pertinent to consideration of legislative or oversight action. A briefing may be conducted by a senator or representative's staff member, a committee staff member, a representative of one of the congressional support agencies, a lobbyist, an individual or group from a federal agency, a concerned constituent, or a representative of the scientific or engineering communities.

Briefing memo: A useful document for members of Congress, generally a page in length. A typical briefing memo might include sections such as "topic," "background," "key issue," and "recommended action or position to take."

Calendar: The list on which all bills are placed as they are reported favorably by committees for House or Senate consideration. The calendar permits an orderly treatment of bills as they are presented to Congress. While there is only one calendar in the Senate, there are a number of calendars in the House of Representatives. Bills may remain on a calendar for the entire legislative session without being called up on the floor for debate, while others are there for only short periods of time.

A calendar of the House of Representatives, together with a history of all measures reported by a standing committee of either house, is printed each day the House is sitting. This information is made widely available for all who are interested. As soon as a bill (or other form of congressional action) is favorably reported, it is assigned a calendar number on either the union calendar or the house calendar, the two principal calendars of business in the House. The calendar number is printed on the first page of the bill.

> **Committee calendar:** Each committee in the House and Senate periodically prepares an updated listing of the status of each piece of legislation that has been referred to the committee. This document is called the committee calendar.

> **Consent calendar:** If a measure pending before the House of Representatives is of a noncontroversial nature, it may be placed on the so-called consent calendar. On the first and third Monday of each month the speaker of the House directs the clerk to call the bills that have been on the consent calendar for three days in the order of their appearance on that calendar. If objection is made to any bill, it is carried over on the calendar to the next day when the consent calendar is again called. If objection is made during this period by three or more members, it is stricken from the calendar and may not be placed on the consent calendar again in the same session of Congress. If no objection is heard and the bill is not "passed over" by request, it is passed by unanimous consent without debate. Ordinarily, the only amendments considered are those sponsored by the committee that reported the bill.

> To avoid the passage without debate of measures that may be controversial or are sufficiently important or complex to require full discussion and debate, there are six official objectors — three majority and three minority — who make a careful study of bills

on the consent calendar. Objection to a bill that leads to its elimination from the consent calendar does not necessarily mean final defeat of the bill since it may then be brought up for consideration in the same way as any other bill on the House or union calendar in the House.

Private calendar: Bills that affect an individual rather than the population at large. A private bill is used for legislative relief in matters such as immigration and naturalization and claims by or against the United States. A private bill is referred to the private calendar; this calendar is called up in the House on the first and third Tuesdays of each month.

Union calendar: Bills are referred to the calendar of the Committee of the Whole House on the State of the Union (commonly known as the union calendar) if they raise revenue, generally appropriate funds, or are of a public character and directly or indirectly appropriate money or property. The great majority of public bills and resolutions in the House are placed on this calendar.

Wednesday calendar: Another means for the House to dispense with legislation quickly. Each Wednesday, the Clerk calls the names of standing committees in alphabetical order. When called, a committee may raise for consideration any bill it reported on the previous day and pending on either the House or the union calendar. Not more than two hours of general debate is permitted on the measure. The affirmative vote of a simple majority is sufficient to pass the bill.

Caucus: A group of persons, belonging to the same political party or faction, who meet to decide on policies and/or candidates.

Cloakroom: The anteroom to the House or the Senate chamber. In the old days, this literally referred to the place where members would hang their cloaks.

Cloture: A vote of two-thirds of the members of a house present needed to stop a filibuster and bring a bill to vote.

Committee: The House and Senate form committees to prepare legislation for action or to make investigations. Most standing committees are divided into subcommittees that study legislature, hold hearings, and report bills to the full committee, which can them report the legislation for action by the House and Senate.

Concurrent resolution: A matter affecting the operations of both houses of Congress is usually initiated in the form of a concurrent resolution. These are not customarily matters of legislative character but are merely expressions of fact, principle, opinion, or purpose of the two houses.

Congressional Budget and Impoundment Control Act of 1974: Commonly called the "Congressional Budget Act," this statute revolutionized the congressional budget process. The Congressional Budget Act was the result of a two-year study by Congress of the procedures that should be adopted for the purpose of improving congressional control of budget outlay and receipt.

Congressional budget cycle: Title III of the Congressional Budget Act establishes a timetable for various phases of the congressional budget process and prescribes the actions to be taken at each point.

Congressional Budget Office (CBO): A nonpartisan agency designed to provide Congress with information needed to make informed decisions about budget policy and national priorities. The CBO monitors the economy and estimates the impact on the economy of government actions; improves the flow and quality of budget information; and analyzes the costs and effects of alternative budget choices.

Congressional Record: The proceedings of the House and the Senate are printed and published daily in the *Congressional Record*, which is mailed to subscribers, postage free, for $37.50 for six months, $75 per year, or $0.50 per copy payable in advance.

Congressional Research Service (CRS): Established in 1914 as the Legislative Reference Service, to provide members of Congress, committees, and staff with information in an objective, nonpartisan, and scholarly manner. Services include analysis of issues before Congress, legal research and analysis, consultation with members of Congress, assistance with statements and speech drafts, and general reference assistance. CRS is a component of the Library of Congress.

Constituent: Every citizen, with the exception of those in the District of Columbia, is represented in Washington by one member of the House of Representatives and two senators to whom the constituent may turn for assistance or with ideas for legislation.

District work period: A regularly scheduled period when the Senate or the House of Representatives is not in session (see appendix).

Document room: The same night that a bill is introduced in either the House or the Senate, it is printed and is available the next day from the document room of the appropriate house. Similarly, committee reports and public laws may be obtained, free of charge, from these document rooms. In order to get a document, send a self-addressed label and a note specifying the document by its appropriate prefix (e.g., H.R. 703) to

Superintendent	Superintendent
Senate Document Room	House Document Room
S-325 Capitol	H-226 Capitol
Washington, D.C. 20510	Washington, D.C. 20515

Documents may be obtained in person by visiting these rooms located in the main Capitol Building.

Engrossed bill: Once a bill has passed the House or the Senate, a copy is printed on blue paper with all the amendments in place. This engrossed bill is delivered in a rather formal ceremony to the other house while the body is actually sitting. Concurrence of the other house is thereby sought on the legislative matter.

Expiring legislation: In November of each year, congressional staff examine the list of legislative items that expire in the following calendar year. This list is useful in gathering forces to lobby effectively for changes in the legislation. Expiring legislation is often the subject of committee and subcommittee hearings early in the next congressional year.

Federal Register: A document published daily that includes federal agency regulations, proposed regulations, and other legal documents of the executive branch.

Filibuster: A delaying tactic used by senators to prevent a vote. Senator Strom Thurmond established the record for filibustering when he spoke against a civil rights bill in 1957 for 24 hours and 18 minutes. The technique may be used by a single dissident or a group of dissidents who have banded together to obstruct the passage of a bill. A filibuster may be stopped only by invoking cloture.

Fiscal year: The period from October 1 to September 30 of government financial operations. Government-funded programs are authorized and funds appropriated coinciding with the fiscal year rather than the calendar year.

Gallery: The balcony from which the public may observe the workings of the House or the Senate when they are in session.

General Accounting Office (GAO): Headed by the Comptroller General of the United States, the General Accounting Office reviews and evaluates government programs carried on under existing law. This includes general review, evaluation, analysis, and audit functions.

Hearing: If a bill is of sufficient importance, public hearings are scheduled. Announcements of hearings may be found in the Daily Digest portion of the *Congressional Record*. Personal notice, usually in the form of a letter, but possibly in the form of a subpoena, is sent to individuals, organizations, and government departments and agencies that would be affected by the bill. Any interested party may ask to be heard. However, minor witnesses may be requested merely to submit written statements for the hearing record rather than to testify orally. Unless otherwise specified, hearings are open to the public. The product of a hearing, whether it is conducted by a committee or a subcommittee, is a transcript published as a separate, usually bound, document available for distribution.

Hill, The: Congress and its associated agencies. Short for "Capitol Hill."

H.R.: The letters affixed to a bill introduced to the House of Representatives.

Introduction of a bill: In the House of Representatives, a member simply drops a typed bill into what is called the hopper, thereby indicating its introduction. The member is not required to ask permission to introduce the measure or to make any statement at the time of its introduction. In the Senate, the procedure is more formal. At the time reserved for such purposes, the senator who wishes to introduce the measure rises and states that a bill is being offered for introduction. The bill is then sent by page to the secretary's desk.

Introductory remarks: The sponsoring senator or representative may ask permission to read or have printed in the *Congressional Record* remarks made at the time of a bill's introduction. These introductory remarks often contain statistics or other descriptive illustrations of need for the legislation. The scientist who has conducted a national survey or who has information as a constituent pertinent to the remarks of the bill's sponsor may find his or her facts or figures incorporated in the introductory remarks of the member of Congress.

Joint resolution: Joint resolutions may originate either in the House of Representatives or in the Senate, not jointly in both houses. There is little practical difference between a bill and a joint resolution, although there is less of the latter. Joint resolutions may represent amendments to the Constitution, which must be approved by two-thirds of both houses and then sent to the states for their ratification.

Legislative counsel: A bill must state in precise legal language what it does and does not do. Hence, bill drafting has come to require a high degree of technical writing skill not possessed by the average citizen.

Most members of Congress call upon specialists to write legislation. After a senator or representative conceives the idea for a bill, the proposal is sketched out in rough draft and sent to the legislative counsel's office for transformation into legal language. Both the House and the Senate have their own legislative drafting staffs.

Lobby: To exert influence on a member of Congress to vote for or introduce legislation desired by an interest group. Also, an interest that engages in such activity.

Majority leader: In the House of Representatives, the majority leader serves as the speaker's chief lieutenant, handling partisan affairs. In the Senate, the majority leader is the chief Senate officer in practice, determining the Senate's legislative schedule and often leading the majority side in legislative debate.

Majority whip: In the House the majority whip is third in command and is primarily responsible for polling the members of the party on pending legislation, informing them of bills and schedules, and summoning them to the floor for crucial votes. In the Senate, the majority whip fills in during the majority leader's absence from the Senate floor.

Mark-up: Line-by-line, highly technical consideration of a bill by subcommittee or committee after hearings are completed. During the mark-up, views of both proponents and opponents of the legislation are studied in detail, legislative language is perfected, and a vote is taken of subcommittee or committee members to determine whether the bill should be reported favorably, reported unfavorably, or tabled. Amendments to the legislation by subcommittee or committee members are also considered during the mark-up.

Minority leader: In the House, the minority leader is the highest ranking member of the minority party; his or her duties include lining up opposition to legislation proposed by the majority party. In the Senate, the minority leader plays a similar role and often confers with the majority leader to obtain prior agreement on legislative scheduling, parliamentary tactics, and the like.

Minority whip: In both the House and the Senate, the minority whip assists the minority leader in polling members on pending legislation, informing them of bills and schedules, and summoning them for crucial votes.

Office of Technology Assessment (OTA): The Office of Technology Assessment is a congressional support agency. Its purpose is to help legislators anticipate the consequences of technological change. OTA is overseen by a nonpartisan congressional board comprised of six senators and six members of the house. Through its professional staff, OTA responds to requests for studies suggested by the board.

Opening remarks: The statement made by the chairperson of a committee or a subcommittee at the time of hearings. The opening remarks lay out the goal of the hearings as well as amplify on the content of the legislation ultimately considered by the members of the committee or the subcommittee. As in the case of introductory remarks, opening remarks may utilize information provided by experts to document the need for a public hearing.

Oversight: The act of reviewing and monitoring federal programs and policies. Oversight responsibilities typically belong to standing committees in various jurisdictional areas in the House and in the Senate.

Pocket veto: The President has 10 days in which to sign a bill that has been passed by both the House and the Senate. If he signs it, it becomes law. If he does not approve it, he can veto it and send it back to the originating house. If the President does not wish to sign the bill or to veto it, and the Congress adjourns before the 10-day period elapses, the bill fails to become law through a "pocket veto."

P.L.: Letters signifying a public law. A bill that becomes a public law is assigned the letters P.L. and given a number. The first two digits indicate the number of the Congress in which the law was enacted. Copies of laws are available by number from the document room of either house.

President of the Senate: The Vice President of the United States presides over the Senate but is not a member of any standing committees and has minimal powers. The President of the Senate usually presides only during important debates.

President pro tempore: The senior member of the majority party in the Senate is usually elected as President *pro tempore*. The President *pro tempore* presides over the Senate in the Vice-President's absence.

Rescission: Literally, a taking away. The act of canceling or voiding, usually of appropriations of government funds.

Report: An official document of a Senate or House committee or subcommittee that summarizes the findings and actions taken by the committee or subcommittee pursuant to hearings, mark-up, and committee discussion of a bill. Reports referred to the House or to the Senate for consideration are given numbers much like numbers given to bills and may be obtained through House or Senate document offices in the manner described elsewhere.

Representative: Individuals elected to membership in the U.S. Congress for a period of two years, representing a district of approximately 500,000 citizens. A representative must be at least 25 years of age, have been a citizen of the United States for seven years, and, when elected, have been a resident of the state in which he or she is chosen.

Roll call: Casting votes in the Senate or in the House of Representatives by calling the name of each member. Roll call votes are published in the *Congressional Record* on the day after the vote.

Rule: An ordinary bill reported out by committee and assigned to a calendar in the House of Representatives must still clear the Committee on Rules before it reaches the House floor. The Committee on Rules was established to determine which bills deserve to proceed and what ground rules will be used for debate. A bill that has been cleared by the rules committee is accompanied by a resolution that specifies the rules for debating the bill. The resolution itself is debated for up to one hour before the bill is brought up. Rules vary, but some typical ones include a rule that says a bill will be debated for two hours after which the bill is open to amendments, or a rule that says a controversial bill may have from 8 to 10 hours for general debate.

S.: A bill introduced in the Senate is designated by the letter "S." followed by a number.

Senator: One of two individuals elected from each state of the union to serve in the U.S. Congress for a period of six years. A senator must be at least 30 years old, have been a citizen of the United States for nine years, and, when elected, must be a resident of the state from which he or she is chosen.

Session: A calendar year. Congress convenes for two years, the first year of which is referred to as the "first session" and the second year, the "second session."

Speaker: The chief of the House of Representatives. The speaker is chosen by the majority party in a presession caucus and is formally nominated along with the minority party candidate, as the first order of business in the first session of a new Congress. The speaker is the leader of the majority party but is expected to remain impartial in rulings while presiding over the House. The speaker has broad powers over legislative scheduling, committee appointments, rules during debate, and recognition of members from the floor.

Sponsorship: A bill introduced in the House of Representatives may have no more than 25 members signing as cosponsors. In the Senate, sponsorship is unlimited. Occasionally, a member may insert the words "by request" after his or her name to indicate that cosponsorship is in compliance with the suggestion of some other member, or the administration, and does not necessarily reflect the commitment of the sponsor.

Staff: The 100 senators in Congress are served by 7,200 professional and clerical assistants. The 435 representatives in the House are served by 12,300 assistants (in Washington and home states and districts). Staff frame the questions for their bosses' consideration, schedule committee hearings, select witnesses, meet with lobbyists and interested constituents to discuss legislation, and respond to enormous volumes of constituent mail.

State delegation: The total number of representatives and senators elected from a state.

Statutes at large: Existing legislation.

Table: A parliamentary motion to remove a bill from consideration. A bill or legislative proposal may be tabled by a subcommittee or committee for the remainder of a congressional session.

Testify: To make a formal statement in a committee or subcommittee hearing in favor of or opposition to a legislative measure. Individuals invited to testify before a committee or subcommittee are requested to appear as witnesses in person having filed their written statement in advance. After the presentation of their testimony, witnesses are cross-examined by members of the committee or subcommittee holding the hearing.

Veto: Once a bill has been approved by both houses, it becomes "an enrolled bill" and is sent to the White House for the signature of the President. The President has 10 days in which to sign a bill. If he does not approve it, he can veto it and send it back to the originating house. A two-thirds majority of each house is needed to override a veto, at which time it become law despite the President's veto. (See also **Pocket veto.**)

Whip: See **Majority Whip, Minority Whip.**

APPENDIX

—～

A. Addresses and Telephone Numbers for Key Congressional Committees and Other Offices

I. Senate Committees Concerned With Science and Technology

U.S. Senate
Washington, D.C. 20510

Agriculture, Nutrition, and Forestry
328A Russell Office Building
202/224-2035

Appropriations
S-128 Capitol Building
202/224-3471

> Subcommittee on VA, HUD, and
> Independent Agencies
> 142 Dirksen Senate Office Building
> 202/224-7211

Armed Services
228 Russell Senate Office Building
202/224-3871

> Subcommittee on Defense Industry and Technology
> 228 Russell Senate Office Building
> 202/224-3871

> Subcommittee on Strategic Forces and
> Nuclear Deterrence
> 228 Russell Senate Office Building
> 202/224-3871

Budget
621 Dirksen Senate Office Building
202/224-0642

Commerce, Science, and Transportation
508 Dirksen Senate Office Building
202/224-5115

> Subcommittee on Science, Technology, and Space
> 427 Hart Senate Office Building
> 202/224-9360

Energy and Natural Resources
304 Dirksen Senate Office Building
202/224-4971

> Subcommittee on Energy Research and Development
> 312 Hart Senate Office Building
> 202/224-4971

Environment and Public Works
458 Dirksen Senate Office Building
202/224-6176

> Subcommittee on Toxic Substances,
> Environmental Oversight, Research and Development
> 458 Dirksen Senate Office Building
> 202/224-6176

Labor and Human Resources
428 Dirksen Senate Office Building
202/224-5375

II. House of Representatives Committees Concerned With Science and Technology

U.S. House of Representatives
Washington, D.C. 20515

Agriculture
1301 Longworth Office Building
202/225-2171

Appropriations
218 Capitol Building
202/225-2771

Subcommittee on Defense
H-144 Capitol Building
202/225-2847

Subcommittee on VA, HUD, and
Independent Agencies
H-143 Capitol Building
202/225-3241

Armed Services
2120 Rayburn House Office Building
202/225-4151

Subcommittee on Research and Development
2120 Rayburn House Office Building
202/225-4140

Budget
214 O'Neill House Office Building
202/226-7200

Education and Labor
2181 Rayburn House Office Building
202/225-4527

Energy and Commerce
2125 Rayburn House Office Building
202/225-2927

Subcommittee on Health and the Environment
2415 Rayburn House Office Building
202/225-4952

Subcommittee on Energy and Power
331 Ford House Office Building
202/226-2500

Subcommittee on Transportation and Hazardous Materials
324 Ford House Office Building
202/225-9304

Merchant Marine and Fisheries
1334 Longworth House Office Building
202/225-4047

Science, Space, and Technology
2320 Rayburn House Office Building
202/225-6371

> Subcommittee on the Environment
> 388 Ford House Office Building
> 202/226-6980
>
> Subcommittee on Energy
> B374 Rayburn House Office Building
> 202/225-8056
>
> Subcommittee on Investigations and Oversight
> 822 O'Neill House Office Building
> 202/225-4494
>
> Subcommittee on Space
> 2320 Rayburn House Office Building
> 202/225-7858
>
> Subcommittee on Technology and Competitiveness
> B374 Rayburn House Office Building
> 202/225-9662
>
> Subcommittee on Science
> 2319 Rayburn House Office Building
> 202/225-8844

NOTE: Any congressional office can be reached through the Capitol switchboard: 202/224-3121.

— ‿

III. Congressional Support Agencies Concerned With Science and Technology

Congressional Budget Office
Fourth Floor
Ford House Office Building
Washington, D.C. 20515
202/226-2621

Congressional Research Service
Library of Congress
First Street and Independence Avenue, S.E.
Washington, D.C. 20540
202/707-5700

Environment and Natural Resources Policy Division
423 Madison Building
202/707-7232

Science Policy Research Division
413 Madison Building
202/707-7040

General Accounting Office
441 G Street, N.W.
Washington, D.C. 20548
202/275-2812

Resources, Community and Economic
Development Division
202/275-3567

Office of Technology Assessment
600 Pennsylvania Avenue, S.E.
Washington, D.C. 20003
202/224-8996

Energy, Materials, and International Security Division
202/228-6750

Health and Life Sciences Division
202/228-6500

Science, Information and Natural Resources Division
202/228-6750

IV. Congressional Leadership Organizations Concerned With Science and Technology

Congressional Biotechnology Caucus
2241 Rayburn House Office Building
Washington, DC 20515
202/225-2815

Congressional Caucus for Science and Technology
1717 Longworth House Office Building
Washington, DC 20515
202/225-5425

Congressional Clearinghouse on the Future
555 Ford House Office Building
Washington, DC 20515
202/226-3434

Congressional Competitiveness Caucus
706 Hart Senate Office Building
Washington, DC 20510
202/224-2651

Congressional Space Caucus
2350 Rayburn House Office Building
Washington, DC 20515
202/225-2631

Democratic Energy Policy Task Force
364 Dirksen Senate Office Building
Washington, DC 20510
202/224-4971

Environmental and Energy Study Conference
515 Ford House Office Building
Washington, DC 20515
202/226-3300

B. Suggested Readings

I. The Governmental Environment:
A Focus on the Congress

Burns, James M.; Peltason, J.W.; and Cronin, Thomas E. *Government by the People*. 13th ed. Englewood Cliffs, NJ: Prentice-Hall, 1989.

Carnegie Commission on Science, Technology, and Government. *Science, Technology, and Congress: Analysis and Advice from the Congressional Support Agencies*. New York, 1991.

Carnegie Commission on Science, Technology, and Government. *Science, Technology, and Congress: Expert Advice and the Decision-Making Process*. New York, 1991.

Chubb, John E., and Peterson, Paul E. *Can the Government Govern?* Washington: Brookings Institution, 1989.

Congressional Quarterly. *How Congress Works*. Washington: CQ Press, 1983.

Duncan, Phil, ed. *Congressional Quarterly's Politics in America: 1992, the 102nd Congress*. Washington: CQ Press, 1991.

Davidson, Roger H., and Oleszek, Walter J. *Congress and Its Members.* 3d ed. Washington: CQ Press, 1990.

Dodd, Lawrence C., and Oppenheimer, Bruce I. *Congress Reconsidered.* 4th ed. Washington: CQ Press, 1989.

Dye, Thomas R., and Zeigler, L. Harmon. *The Irony of Democracy: An Uncommon Introduction to American Politics.* 7th ed. Monterey, CA: Brooks/Cole, 1987.

Fisher, Louis. *The Politics of Shared Power.* Washington: CQ Press, 1987.

Keefe, William J. *Congress and the American People.* 3d ed. Englewood Cliffs, NJ: Prentice Hall, 1988.

Oleszek, Walter J. *Congressional Procedures and the Policy Process.* 3d ed. Washington: CQ Press, 1989.

Smith, Steven S., and Deering, Christopher J. *Committees in Congress.* 2d ed. Washington: CQ Press, 1990.

Wittenberg, Ernest, and Wittenberg, Elisabeth. *How to Win in Washington.* Cambridge, MA: Basil Blackwell, 1989.

Woods, Patricia D. *The Dynamics of Congress: A Guide to the People and Process in the U.S. Congress.* 3d ed. Washington: Woods Institute, 1987.

II. Preparing and Giving Testimony and Presentations:

Arrendo, Lani. *How to Present Like a Pro: Getting People to See Things Your Way.* New York: McGraw-Hill, 1991.

Bedrosian, Margaret McAuliffe. *Speak Like a Pro in Business and Public Speaking.* New York: Wiley, 1987.

Bell, Arthur H., and Skopek, Eric W. *The Speaker's Edge: Tips for Confident Presenting.* Westbury, NY: Asher-Gallant Press, 1988.

Boylan, Bob. *What's Your Point? A Proven Method for Giving Crystal Clear Presentations!* Wayzata, MN: Point Publications, 1987.

Brooks, William T. *High Impact Public Speaking.* Englewood Cliffs, NJ: Prentice Hall, 1988.

Echeverria, Ellen W. *Speaking on Issues: An Introduction to Public Communication.* New York: Holt, Rinehart, and Winston, 1987.

Gard, Grant C. *The Art of Confident Public Speaking.* Englewood Cliffs, NJ: Prentice Hall, 1986.

Harvey, Thomas A. "Congressional Testimony: A Practical Guide for Newcomers to the Congressional Hearing Process." *Federal Bar News & Journal* 35 (Nov. 1988): 414-15

Humes, James C. *Standing Ovation: How to Be an Effective Speaker and Communicator.* New York: Harper & Row, 1988.

Kenny, Peter. *A Handbook of Public Speaking for Scientists and Engineers.* Bristol: A. Hilger, 1982.

Kougl, Kathleen M. *Primer for Public Speaking.* New York: Harper & Row, 1988.

Leeds, Dorothy. *PowerSpeak: The Complete Guide to Persuasive Public Speaking and Presenting.* New York: Prentice Hall, 1988.

Lustberg, Arch. *Testifying With Impact.* Washington: Association Division, U.S. Chamber of Commerce, 1982.

Minninger, Joan, and Goulter, Barbara. *The Perfect Presentation.* New York: Doubleday, 1991.

Osgood, Charles. *Osgood on Speaking: How to Think on Your Feet Without Falling on Your Face.* New York: Morrow, 1988.

Smith, Terry C. *Making Successful Presentations: A Self-Teaching Guide.* New York: Wiley, 1991.

Vasile, Albert J., and Mintz, Harold K. *Speak With Confidence: A Practical Guide.* 5th ed. Glenview, IL: Scott, Foresman, 1989.

C. Sources of Information About Congress

I. Reference Books

Almanac of American Politics. Profiles of senators, representatives, and governors, with voting records, election results, and district information; published annually by the National Journal, 1730 M Street, NW, Washington, DC 20036; 202/857-1400; $56.95 hardcover, $44.95 softcover, plus 10 percent shipping; available in some bookstores.

Congressional Staff Directory. Comprehensive reference to key congressional staffers, including biographies; published bi-annually by Staff Directories, Ltd., Mount Vernon, VA 22121-0062; 703/739-0900; $59; also available from Trover Book Stores, in Washington, DC.

Congressional Yellow Book. Comprehensive reference guide; published quarterly by Monitor Publishing Co., 1301 Pennsylvania Avenue, NW, Suite 925, Washington, DC 20004; 202/347-7757; annual subscription rate is $185; not available in bookstores.

Politics in America. State-by-state guide, with both national and state profiles and statistics; published annually by Congressional Quarterly Press, 1414 22nd Street, NW, Washington, DC 20037; 202/887-8500; $54.95 hardcover; $39.95 softcover; available in some bookstores.

Senate Telephone Directory. Listing of senators and personal and committee staff; published semi-annually; available from the Superintendent of Documents at the Government Printing Office, Washington, DC 20402; 202/783-3238; $10 (price subject to change).

U.S. House of Representatives Telephone Directory. Listings of members of Congress and personal and committee staff; updated two or three times a year; available from the Superintendent of Documents at the Government Printing Office, Washington, DC 20402; 202/783-3238; $17 (price subject to change).

— —

II. Periodicals

American Caucus. Bi-weekly newspaper (for more general audiences) that provides information about issues coming before Congress; published by Congressional Quarterly Inc., 1414 22nd Street, NW, Washington, DC 20037; 202/887-8679 or 1-800-544-0155; subscription rate is $39 per year (discounts for group orders).

Aviation Week & Space Technology. Weekly news magazine covering national security, arms control, space, and commercial aviation issues; available from McGraw Hill, Inc., 1221 Avenue of the Americas, New York, NY 10020; 212/512- 3288; subscription rate is $82 per year.

Chemical & Engineering News. Weekly magazine; published by the American Chemical Society, 1155 16th Street, NW, Washington, DC 20036; 202/872-4363; subscription rate is $105 per year (regardless of membership).

Congressional Insight. Weekly newsletter analyzing the people and politics that shape Capitol Hill decisions; published by Congressional Quarterly Inc., 1414 22nd Street, NW, Washington, DC 20037; 202/887-8500 or 1-800-854-9043; subscription rate is $299 per year.

Congressional Quarterly. Weekly publication that covers major issues coming before Congress (including votes and status of legislation); published by Congressional Quarterly Inc., 1414 22nd Street, NW, Washington, DC 20037; 202/887-8500 or 1-800-854-9043; subscription rate is $1,220 per year.

Environmental and Energy Study Conference Weekly Bulletin. Weekly publication providing background, history, and overview of debate on environmental and energy issues before Congress; available from the Environmental and Energy Study Institute, 122 C Street, NW, Suite 700, Washington, DC 20001; 202/628- 1400; introductory subscription rate is $295 per year.

National Journal. Weekly publication with in-depth articles and analysis of politics and government; available from National Journal, 1730 M Street, NW, Suite 1100, Washington, DC 20036; 202/857-1400; subscription rate is $767 per year.

Nature. Weekly British scientific journal; published by Nature Publishing Co., 65 Bleecker Street, New York, NY 10012; 202/737-2355; subscription rate is $135 per year.

Roll Call. Capitol Hill's local newspaper; published by Levitt Communications, 900 Second Street, NE, Suite 107, Washington, DC 20002; 202/289-4900; subscription rate is $185 for 96 issues.

Science. Weekly journal of the American Association for the Advancement of Science; contact AAAS Membership and Circulation, 1333 H Street, NW, Washington, DC 20005; 202/326-6400; subscription is included in $87 annual dues for members ($205 for nonmembers).

Science & Government Report. Bi-monthly newsletter of science policy issues; published by Science & Government Report, Inc., 3736 Kanawha Street, NW, Washington, DC 20015; 202/244-4135; introductory subscription rate is $148 per year.

— ⬝

D. Obtaining Congressional Documents

These contacts should be useful in acquiring congressional documents. Remember to have any specific information about the desired document, such as a document's title, author(s), or publication number, ready when calling or visiting. Also, remember that when calling, it is not unusual to be put on hold for a number of minutes before speaking with someone.

Superintendent of Documents
Senate Document Room
B04 Hart Senate Office Building
Washington, DC 20510
202/224-7860

> No publication list is available. You may order up to six different documents at no charge. Multiple copies of documents are available from the Government Printing Office for a fee. When ordering, have report titles and public law numbers ready.

Superintendent of Documents
House Document Room
B-18 Ford House Office Building
Washington, DC 20515
202/225-3456

> No publication list is available. You may order up to six different documents at no charge. Multiple copies of documents are available from the Government Printing Office for a fee. When ordering, have report titles and public law numbers ready.

Office of Technology Assessment (OTA)
600 Pennsylvania Avenue, SE
Washington, DC 20003
202/224-8996

> Publication list may be requested free of charge by phone or in writing. There is a charge for most publications. When ordering, have report titles ready.

General Accounting Office (GAO)
Document Handling and
Information Services Facility
700 4th Street, NW
Washington, DC 20548
202/275-6241

The GAO publishes a monthly list of its reports. While the first copy of a report is available at no cost, there is a charge for additional copies. You should know title of report to be requested.

Superintendent of Documents
U. S. Government Printing Office (GPO)
Washington, DC 20402
202/783-3238

Call before requesting by mail to be sure that the document is in stock. Orders should include title of publication and GPO stock number if possible.

In person: 710 N. Capitol Street, NW
Washington, DC 20401
202/275-2091
or
1510 H Street, NW
Washington, DC 20005

A check or money order made out to the Superintendent of Documents should accompany orders. Orders may also be charged to a prepaid Superintendent of Documents deposit account with Visa or Mastercard. The GPO also handles subscriptions to the *Congressional Record* and the *Federal Register*, as well as a Monthly Catalog of U. S. Government Publications. At no cost, one may request *U.S. Government Books, New Books List,* and *Government Periodicals and Subscription Services.*

Congressional Budget Office
Ford House Office Building, 4th Floor
Washington, DC 20515
202/226-2809

A list of publications can be requested by phone at no cost. In person: Second & D Streets, SW, 4th Floor. While the first copy of a report is available at no cost, there is a charge for additional copies.

Congressional Research Service (CRS)
Library of Congress
First and Independence Avenues, SE
Washington, DC 20540
202/707-5700

In general, CRS reports are available only through congressional offices.

— —

E. The Library of Congress

Serving as the nation's main repository, the Library of Congress is an invaluable source of information. The Library maintains collections of journals, newspapers, manuscripts, films, photos, maps, presidential papers, and recorded speeches as well as books. The public may use the Library of Congress as a general reference library.

Library of Congress
First and Independence Avenues, SE
Washington, DC 20540

General information: 202/707-5000

Recorded information: 202/707-6400

Reference Services: 202/707-5640
Reference services are available by phone, by mail, or in person. There are various restrictions and time constraints (e.g., some materials are off-site and require several days to acquire).

Loan Division: 202/707-5441
Note: All Library of Congress materials are for "in library" use; however, some materials may be obtained through interlibrary loans.

Photo duplication Services: 202/707-5640
Note: There is a fee for photo duplication services, and they are subject to restrictions.

Reading Rooms:
Government Publications Reading Room: 202/707-5647
Open 8:30 a.m. - 9:30 p.m. (M-F);
8:30 a.m. - 5 p.m. (Sat); 1 - 5 (Sun).
Location: Room 133, Madison Building

Library of Law: 202/707-5079
Open 8:30 a.m. - 9:30 p.m. (Mon. - Fri.);
8:30 a.m. - 5 p.m. (Sat); 1 - 5 (Sun).
Location: Room 201, Madison Building

Main Reading Room:
Open 8:30 a.m. - 9:30 p.m. (Mon - Fri);
8:30 a.m. - 5 p.m. (Sat); 1 - 5 (Sun).
Location: Jefferson Building (2nd Street entrance)

Periodicals Reading Room: 202/707-5690
Open 8:30 a.m. - 9:30 p.m. (M-F);
8:30 a.m. - 5 p.m. (Sat); 1 - 5 (Sun).
Location: Room 133, Madison Building

Science Reading Room: 202/707-5639
Open 8:30 a.m. - 9:30 p.m. (Mon. - Fri);
8:30 a.m. - 5 p.m. (Sat.); 1 - 5 (Sun).
Location: 5th Floor, Adams Building

F. Washington Offices of Professional Societies and Other Relevant Organizations

I. Professional Societies

American Association for the Advancement of Science
1333 H Street, NW, Washington, DC 20005
202/326-6600
Contact: Albert H. Teich, Director, Science and Policy Programs

American Association of Engineering Societies
1111 19th Street, NW, Suite 608, Washington, DC 20036
202/296-2237
Contact: Mitchell Bradley, Executive Director

American Astronomical Society
1630 Connecticut Avenue, NW, Suite 200, Washington, DC 20009
202/328-2010
Contact: Peter B. Boyce, Executive Officer

American Chemical Society
1155 16th Street, NW, Washington, DC 20036
202/872-4477
Contact: Kathleen A. Ream, Head, Department of Government Relations and Science Policy

American Geological Institute
4420 King Street, Alexandria, VA 22302
703/379-2480
Contact: Marcus E. Milling, Executive Director

American Geophysical Union
1630 Connecticut Avenue, NW, Washington, DC 20009
202/462-6903
Contact: Fred Spilhaus, Director

American Institute of Aeronautics and Astronautics
370 L'Enfant Promenade, SW, Washington, DC 20024
202/646-7404
Contact: Joanne Padron, Deputy Director, Science & Technology
Policy

American Institute of Biological Sciences
730 11th Street, NW, Washington, DC 20001
202/628-1500
Contact: Charles Chambers, Executive Director

American Institute of Physics
1630 Connecticut Avenue, NW, Suite 750, Washington, DC 20009
202/332-9661
Contact: Richard M. Jones, Senior Liaison, Government and Institutional Relations

American Meteorological Society
1701 K Street, NW, Suite 300, Washington, DC 20006
202/466-6070
Contact: Yale Schiffman, Development Director

American Physical Society
529 14th Street, NW, Suite 1050, Washington, DC 20045
202/662-8700
Contact: Robert Park, Executive Director, Washington Office

American Psychological Association
750 First Street, NE, Washington, DC 20002-4242
202/336-6062
Contact: Brian L. Wilcox, Director, Public Policy Office

American Psychological Society
1010 Vermont Avenue, NW, Suite 1100, Washington, DC 20005-4907
202/783-2077
Contact: Alan Kraut, Executive Director

American Society for Agronomy
677 South Segoe Road, Madison, WI 53711
608/273-8080
Contact: Robert F. Barnes, Executive Vice President

American Society for Horticultural Science
113 South West Street, Alexandria, VA 22314-2824
703/836-4606
Contact: Skip McAfee, Executive Director

American Society for Microbiology
1325 Massachusetts Avenue, NW, Washington, DC 20005
202/737-3600
Contact: Robert Watkins, Director of Public Affairs

American Society of Mechanical Engineers
1828 L Street, NW, Suite 906, Washington, DC 20036-5104
202/785-3756
Contact: Philip Hamilton, Managing Director of Public Affairs

American Veterinary Medical Association
1101 Vermont Avenue, NW, Suite 710, Washington, DC 20005
202/789-0007
Contact: Marcia Brody, Senior Policy Specialist

Consortium of Social Science Associations
1522 K Street, NW, Suite 836, Washington, DC 20005
202/842-3525
Contact: Howard Silver, Executive Director

Council of Professional Associations on Federal Statistics
1429 Duke Street, Suite 402, Alexandria, VA 22314
703/836-0404
Contact: Edward Spar, Executive Director

Federation of American Societies for Experimental Biology
9650 Rockville Pike, Bethesda, MD 20814
301/530-7000
Contact: Gar Kaganowich, Director of Public Affairs

Federation of Behavioral, Psychological, and Cognitive Sciences
750 First Street, NE, Suite 5004, Washington, DC 20002-4242
202/336-5920
Contact: David Johnson, Executive Director

Institute of Electrical and Electronics Engineers
1828 L Street, NW, Suite 1202, Washington, DC 20036-5104
202/785-0017
Contact: Leo C. Fanning, Staff Director, Professional Activities

Joint Policy Board for Mathematics
1529 18th Street, NW, Washington, DC 20036
202/234-9570
Contact: Lisa Thompson, Congressional Liaison

National Society for Professional Engineers
1420 King Street, Alexandria, VA 22314-2715
703/684-2873
Contact: Bob Reeg, Manager, Congress and State Relations

Society for Neuroscience
11 Dupont Circle, NW, Suite 500, Washington, DC 20036
202/462-6688
Contact: Betty Willis, Director of Government Affairs

Society for Research in Child Development
(Federal affairs are handled by the American Psychological Society —
see above)

II. Other Organizations

American Association of State Colleges
and Universities (AASCU)
One Dupont Circle, NW
Washington, DC 20036
202/293-7070

A group comprised of chancellors and presidents of state colleges
and universities, it monitors legislation and regulation and provides a forum for exchange of information on issues regarding
higher education.

American Forest Council
1250 Connecticut Avenue, NW
Washington, DC 20036
202/463-2455

Representing wood and paper companies, it provides information
on forest resources and environmental-related industrial policies;
commissions public opinion research on resource and environmental issues.

Association of American Universities (AAU)
One Dupont Circle, Suite 730
Washington, DC 20036
202/466-5030

Represents 58 of the nation's top research universities as well as
four Canadian universities; works closely with NASULGC (see
below).

Association of American Medical Colleges (AAMC)
One Dupont Circle, Suite 200
Washington, DC 20036
202/828-0400

Represents medical colleges on research funding and other issues of interest to the biomedical community.

Conservation Foundation
1250 24th Street, NW
Washington, DC 20037
202/293-4800

Affiliated with the World Wildlife Fund; it researches and writes on environmental issues such as toxic substances control, dispute resolution, resource conservation, and land management.

Council of Graduate Schools in the United States (CGS)
One Dupont Circle, Suite 430
Washington, DC 20036
202/223-3791

Comprised of the deans of the nation's major graduate schools.

Council on Research and Technology (CORETECH)
1735 New York Avenue, NW, Suite 500
Washington, DC 20006
202/628-1700

A group comprised of leaders in education and high-tech industry; its major function is to lobby for increased federal support for R&D and related issues, such as the R&D tax credit.

Council on Competitiveness
900 17th Street, NW, Suite 1050
Washington, DC 20006
202/785-3990

An organization of leaders of industry and higher education concerned with the promotion of policies that will improve U.S. industrial competitiveness. Publishes a variety of reports, including an annual "competitiveness assessment" of the President's proposed budget. A number of scientific and higher education associations (including AAAS) are among its affiliates.

Industrial Research Institute (IRI)
1550 M Street, NW
Washington, DC 20005
202/872-6350

> Comprised of the vice presidents for R&D or their counterparts in
> most of the nation's largest S&T-intensive corporations, IRI holds
> meetings and seminars, conducts studies, and issues publications
> in areas related to industrial R&D.

National Academy of Sciences (NAS)
2101 Constitution Avenue, NW
Washington, DC 20418
202/334-2100

> Congressionally chartered independent organization that promotes
> the use and benefits of science; it advises the federal government
> on science and technology issues. The National Academy of En-
> gineering and the Institute of Medicine are its affiliates; and the
> National Research Council is its operating arm.

**National Association of State Universities and Land Grant Colleges
(NASULGC)**
One Dupont Circle, Suite 710
Washington, DC 20036
202/778-0818

> Represents state universities and land-grant colleges; active in
> many areas, including student aid, health, and agriculture; works
> closely with AAU.

Research!America
99 Canal Center Plaza, Suite 250
Alexandria, VA 22314
703/739-2577

> An organization intended to build public support and apprecia-
> tion for health research; operates mainly through advertisements
> and public service announcements in print and electronic media.

Other Organizations

Other organizations active in science and technology issues include
groups that lobby for increased funding in particular issue areas, such
as the Function 250 Coalition (which is comprised mainly of scientific and higher education organizations and is concerned with agen-

cies and programs in federal budget function category 250, such as NSF and NASA) and the Delegation for Basic Biomedical Research. There are also a wide variety of trade and industry associations in such fields as aerospace, computers, and electronics, and numerous public interest groups concerned with science-related issues, especially environment, health, and consumer protection. Congressional Quarterly's *Washington Information Directory* (published annually) is a good source of information, including addresses and telephone numbers for many of these organizations.

G. The Congressional Year

The House and Senate follow similar but not identical calendars. Each new Congress convenes on January 3, following the November election, as provided in the Twentieth Amendment to the Constitution. Generally both houses then recess until late in January, when they reconvene for the President's State of the Union Address and budget message. The main exception to this occurs during presidential election years, when Congress must convene in a joint session after the election to count the electoral votes for the President and Vice President.

A considerable amount of informal activity occurs between the election and the initial convening date and during the January recess. These activities include election of leadership, committee and subcommittee assignments, and allocation of office space. After Congress reconvenes in late January, legislative work begins (or continues, in even-numbered years). Since congressional committees require some time to prepare legislation, very little legislation is passed in the early weeks of the session.

Congress recesses several times during the year. Most of the dates for recesses vary from year to year; however, they generally follow patterns established by long tradition. Normally, both House and Senate recess for about a week in February around President's Day, in April around Easter, in May for the Memorial Day weekend, in early July around Independence Day, and often for the latter part of August through Labor Day. When reading a House calendar it is important to note that recesses are designated as "district work periods." During presidential election years, both Houses will also recess for the political conventions.

Generally, Congress aims to complete its work by early October. However, in recent nonelection years, because of increasingly fractious debates over budget issues, Congress has seldom completed business much before Christmas. The date for adjournment is voted on by the House and Senate.

H. Legislative Buzzers, Bells, and Signal Lights

All congressional hearing rooms contain signal buzzers or bells and many also have lights. These devices are used to inform members in hearings of things that are taking place on the floor of the House or Senate. Often these floor activities require members to leave a hearing abruptly. Few witnesses or observers in these hearings understand the significance of these often distracting signals, which differ between the two Houses. Following are lists of House and Senate signals:

House Legislative Electric Bell Signals

1 ring — Teller vote. Not recorded

1 long ring (pause, followed by 3 rings) —Signals the start or continuation of a notice quorum call.
(This call will be terminated if and when 100 members appear.)

1 long ring — Termination of a notice quorum call.

2 rings — Electronically recorded vote.

2 rings (pause, followed by 2 rings) — Manual roll call vote — (the bells will be sounded again when the clerk reaches the R's).

2 rings (pause, followed by 5 rings) — First vote under Suspension of the Rules or on clustered votes. (2 rings will be rung 5 minutes later. The first vote will take 15 minutes with successive votes at intervals of not less than 5 minutes. Each successive vote signaled by 5 rings.)

3 rings — Quorum call (either initially or after a notice quorum has been converted to a regular quorum call. The bells are repeated 5 minutes after the first ring. Members have 15 minutes to be recorded.)

3 rings (pause, followed by 3 rings) — Manual quorum call (the bells will be sounded again when the clerk reaches the "R"s).

3 rings (pause, followed by 5 rings) — Quorum call in Committee of the Whole, which may be immediately followed by a 5-minute recorded vote.

4 rings — Adjournment of the House.

5 rings — Five-minute electronically recorded vote.

6 rings — Recess of the House.

12 rings rung at 2 second intervals — Civil defense warning.

Senate Legislative Buzzers and Signal Lights

Presession signals: 1 long ring at hour of convening; 1 red light to remain lighted at all times while Senate is in normal session.

1 ring — Yeas and nays.

2 rings — Quorum call.

3 rings — Call of absentees.

4 rings — Adjournment or recess (end of daily session).

5 rings — Seven and a half minutes remaining on yea or nay vote.

6 rings — Morning business concluded (lights cut off immediately)

Recess during daily session (lights stay on during period of recess).

MAP OF CAPITOL HILL

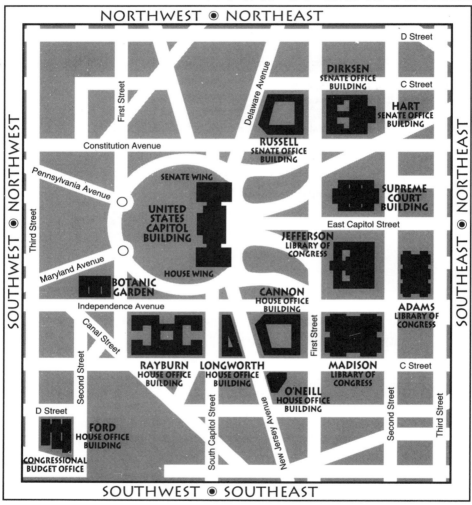

Adapted from *Congressional Yellow Book* Vol. XVIII, No. 2, Summer 1992 (Washington, DC: Monitor Publishing Co.).

INDEX